SOTHIS PRESS

Sothis Press aims to completely reimagine the public domain book with an emphasis on universal access to captivatingly designed print editions. We believe that technology will enable every single book ever published to be perpetually accessible in beautiful print form. Sothis is beauty in ubiquity.

www.sothis.press
1 Waterhouse Square, London, ECIN 2ST

Calcium Carbide and Acetylene

George Gilbert Pond

SOTHIS PRESS

ISBN 978-1-0365-1821-9

Calcium Carbide and Acetylene
is in the public domain.

This edition is published by
 The company
is registered in England and Wales.
The Company Number is 14900389.

THE PENNSYLVANIA STATE COLLEGE

BULLETIN

OF THE

DEPARTMENT OF CHEMISTRY

CALCIUM CARBIDE

AND

ACETYLENE

Revised Edition

BY

GEORGE GILBERT POND, PH. D.
PROFESSOR OF CHEMISTRY

1909

TABLE OF CONTENTS

PREFACE.

The original Bulletin, of which this is a second and revised edition, was issued in 1900, when the industry was quite in its infancy, by the Pennsylvania Department of Agriculture, and known as Bulletin No. 57, entitled "The Application of Acetylene to Country Homes." In the years since it was published the acetylene industry has made such tremendous strides in all civilized countries of the globe that the former publication has become entirely antiquated.

So many and such urgent requests have come to the author, at various times and from all sorts of sources, for a new and revised edition of the Bulletin, as to amount to a practical demand upon his attention, and the present work, which aims to be prepared in the same intelligible and untechnical style as the former one, possibly requires no other explanation of its existence.

It has been thought best to issue this edition as a bulletin of the Department of Chemistry of The Pennsylvania State College, and to give it a more comprehensive title.

The author claims for it only the same fair consideration of all interests involved as was aimed at in the earlier work, and bespeaks for it the same kindly reception.

State College, Pa., January 1, 1909.

CALCIUM CARBIDE AND ACETYLENE.

INTRODUCTION.

An old writer, I do not recall what one, relates that after sunset Nineveh was illuminated, in its streets and palaces, by lights so brilliant that it was difficult to distinguish night from day. In such an account it is doubtless necessary to make ample allowance for exaggeration, yet the fact remains that we have lost all hope of ever knowing by what kind of light Nineveh and Babylon were illuminated. Still, recent discoveries permit us at least to hope that in our towns lighted by acetylene, this artificial sunlight, as we are wont to call it, night shall be no more regretted than day.

For a moment it would be well to glance at the past and review the years as they have advanced. The Greeks and Romans lighted their cities in primitive fashion with torches which were mere pieces of resinous wood, or with metallic tubes containing tow soaked in pitch or resin, or with candles made of tallow, resin or wax. Later came lamps, vessels made of burnt clay or of metal, filled with oil and provided with a wick. The many specimens of lamps which ancient Rome has bequeathed to us present a great variety, often of marvelously graceful forms. The artistic merit of the antique Roman or Grecian lamp is well known, but, naturally enough, neither the beauty nor the richness of the vessel could prevent the flame from being smoky and dim. There were also candelabra, supporting several lamps or flames, as well as lanterns, the walls of which were of translucent horn or of bladder, and in later days of glass, and until comparatively recent times our ancestors possessed no other means of illumination.

Under these conditions, one can readily understand that it was difficult to light city streets at night. Imperial Rome was indeed illuminated on the occasion of certain important festivals. In France, the royal ordinances, which required the inhabitants to maintain a certain number of lamps in each quarter, remained a dead letter till the middle of the seventeenth century. Louis XIV. delighted in adorning his capital; he provided, therefore, for the security of its streets, which were tortuous and notoriously beset with criminals. By letters patent dated August 26, 1662, the King ordered the establishment of movable lights to be placed at the disposal of the public. Torch bearers and lantern carriers were stationed every eight hundred paces, in all quarters of the city, and let themselves to each comer, accompanying him a quarter of an hour for the consideration of a few coppers.

Only a few years later a progressive lieutenant of police compelled the citizens to maintain lanterns during the winter, in each quarter, and thus organized a definite system of lighting. A decree of Parliament, May 23, 1671, fixed the commencement of the season at October 20, and the end at May 31. The seventeenth century saw, before its close, six thousand five hundred lanterns, consuming each night in the streets of Paris, sixteen hundred and twenty-five pounds of candles. Strangers never tired of expressing their admiration of the novel illumination. Dr. Lister, a famous English traveler, writes his impressions, upon his return from Paris in 1688, as follows:

"The streets are lighted all winter, as much when the moon is visible as during the rest of the month. I remark upon it especially, because of the stupid custom which prevails in London of extinguishing the lanterns during half of the month as if the moon must surely be brilliant enough to light the streets, and as if it were impossible for the sky to be clouded in winter. These lanterns are suspended right over the middle of the streets, twenty feet high and a score of paces apart. They are furnished with glass about two feet square and have covers of sheet iron. The cord which suspends them is fastened, under lock and key, to the wall of a neighboring house. In the lanterns are candles sufficient to last till after midnight."

Perfection was far from reached, but in those times people knew how to be contented with little. Illumination, public and private, made no notable advancement before 1764. At this time a man of great intelligence and energy was made chief of police in Paris. He offered a reward to that inventor who should perfect the illumination of the streets. At his suggestion the Academy of Science offered the extraordinary prize of two thousand pounds to the writer who should treat in the most approved manner the subject, "The Best Means of Lighting at Night the Streets of a Great City, Combining Brilliancy, Easy Accomplishment and Economy." Lavoisier presented a memoir and received a gold medal, but the model adopted was that of another who invented an oil street lamp. The apparatus was so remarkable that a prominent statesman of that time wrote to the King, "The light which it gives precludes the thought of any one ever finding anything better." However, something better was found, and in 1821 the Argand chimney was adapted to the street lamp, and increased, to a remarkable degree, the intensity of the flame. Argand was a Genevese chemist, a pupil of the famous Swiss naturalist Saussure. He had completed his university studies in physics and chemistry, and had presented to the Academy a number of remarkable addresses when he invented his lamp with a double current of air, first made in England in 1782. The Argand lamp, however perfect it may have been, had one fault; the level of the oil could not be maintained rigidly constant. A clockmaker, Carçel, solved this problem by attaching to the lower part of the Argand lamp a little force pump driven by clockwork, the office of which was to raise to the wick the oil contained in the lower receptacle. This new lamp acquired for a short time a just notoriety; it was called the Carcel lamp; science adopted the name, and the Carcel became the recognized unit of light. From the Carcel lamp to the Moderator was but a step. Franchot invented the latter by simplifying earlier forms, substituting a spiral spring for the clockwork, and thus came into use a style of lamp employed up to the present day.

During this perfecting of illumination by oil, the candle industry was suffering a transformation. Many investigators had attempted, with little success, to employ the fatty acids in illumination. Meantime the revolution of 1830 was declared. De Milly, a courtier of Charles X., seeing his career suddenly broken off, sought a fortune in this industry, and devoted himself to investigation, which resulted in the greatly improved stearine candle so popular today.

Meantime public lighting had commenced to develop in a new direction. Krueger succeeded, in 1786, in extracting from coal a quantity of gas, sufficient to light the Jesuit College of Stony Hurst in Lancashire, but as this impure gas yielded only a smoky flame, his experiments were discontinued. The same year the chemical laboratory of the University of Würzburg in Germany was lighted with the gas obtained by the dry distillation of bones.

In 1791, Le Bon, a young French engineer, discovered that the gas liberated from calcined wood was ignitible and yielded a brilliant flame. With indefatigable energy he pushed the development of his discovery, and in 1802 accomplished the successful illumination of his house with this gas. But history gives to William Murdoch, the Englishman, the honor of being the true inventor of gas lighting, since he in 1792 lighted his workshop in Cornwall with gas obtained from soft coal. Ten years later Murdoch had established quite extensive gas works near Birmingham, the factory in which James Watt had constructed his steam engines being one of the buildings to be thus illuminated, but it was not till 1812 that gas lighting was adopted to any considerable extent in London. A German named Winsor, who had kept himself acquainted with the results of Le Bon and Murdoch, succeeded in breaking down the powerful prejudice which existed in England against the innovation, lighted the Lyceum Theater with gas in 1803, and in 1810 founded the first gas company in London. From England, Winsor went to France, where Louis XVIII. gave him every facility for exploiting the novel method of illumination. In the face of all sorts of prejudice the industry made steady advances, and the year 1820 saw its introduction to all the larger continental cities.

Oil gas has had a considerable career, and is used to enrich other gases or is burned by itself; it is made by injecting crude oil into hot cylinders or retorts, called the "cracking" process, because the heavy oils of shale or tar origin are thus broken up, or cracked, into light gases.

Water gas, made by the action of steam on red-hot coal, though itself non-luminous, is rendered suitable for illumination by impregnating it with gases derived from oil, a process familiarly known as carburetting. This gas has had an immense effect upon the gas industry of the world, especially in our own country, where either alone, or accompanying the coal gas manufacture, it is known in every large city of the land.

Petroleum, introduced into the markets of the world within the last half century, replaced to a large extent the other oils which had been employed in the earlier forms of lamps, but did not revolutionize the methods of producing light.

Dr. von Welsbach, an Austrian chemist, has outstripped all competitors in solving the problem of securing the most light from little gas. The Welsbach or incandescent mantle is prepared by saturating a cotton fabric with a solution of the rare earths, thoria and ceria. After drying, the vegetable matter is burned away, leaving a fragile skeleton of the oxides, which are rendered incandescent in a non-luminous flame. Placed over a Bunsen flame this mantle produces a brilliant light whereby the candle power of coal gas is increased more than three-fold. In this invention, gas companies possess a most powerful weapon which they are today wielding with great effect in the struggle against electricity.

Electric illumination, whose history is familiar to all of us, has given gas illumination a real competition, though its cost has prevented the destruction of the gas industry.

And now acetylene, a name so long known only to those who frequented the laboratories of science, but to-day on everybody's lips, has spread over all civilized parts of the globe, and the use of this illuminant has made as much progress in a dozen years as its predecessor, coal gas, made in the first half century of its history.*

*NOTE.—Portions of this introduction have been translated and adapted from Capelle's l'Éclairage à l'Acétylène.

DISCOVERY OF CARBIDE OF CALCIUM AND ACETYLENE.

As early as 1836 Edmund Davy, an English chemist, observed that a by-product which he had secured incidentally to the production of metallic potassium, was capable of decomposing water with the evolution of a gas which contained acetylene. In 1862 Woehler, the most famous chemist of his day, announced the discovery of the preparation of acetylene from calcium carbide, which he had made by heating to a very high temperature a mixture of charcoal with some alloy of zinc and calcium. The product could decompose water, like Davy's compound, and yield a gas containing acetylene. Thus the phenomenon which had been observed but not understood by Davy was explained and published by Woehler, and to him belongs the honor of the discovery of calcium carbide and of acetylene. Woehler further pointed out that the new gas burns with a brilliant but very smoky flame, and announced several other characteristics, though it remained for Berthelot, in the same year, to very exhaustively study and describe these new compounds.

Then for nearly thirty years these two substances seem to have been practically forgotten. During all this time acetylene was procurable only by the tedious methods which the early discoverers had used, with the result that even its name was known only to students of science, and it is perfectly safe to say that up to 1892 few even of the professional chemists of the world ever saw an acetylene flame, much less dreamed of it as a commercial possibility.

But with the development of the modern electric furnace, whose high temperature places in the hands of the investigator a means of bringing about chemical changes never before known, the laws which govern its formation could no longer remain undiscovered. It was through the instrumentality of the electric furnace that the possibility of calcium carbide as a commercial product was revealed to Thomas L. Willson, an electrical engineer at Spray, in North Carolina. In May of the above year this gentleman was conducting experiments, with a view to the preparation of metallic calcium, for which purpose he employed an improved Héroult electric furnace with a current of two thousand amperes and thirty-six volts, operating upon a mixture of lime and coal tar. He secured a melted mass of dark color, which on cooling became solid and brittle. Willson is said to have discarded this product, as it was clearly not the material for which he was searching—metallic calcium. It was thrown into a neighboring stream, when to the astonishment of those who saw it, there was suddenly liberated a great quantity of gas which, on being kindled, burned with the now familiar, bright but smoky flame, and gave off quantities of soot which showed that it was not hydrogen, which would be liberated from metallic calcium and water, but a rich hydrocarbon gas. The smelt was repeated, the product submitted to analysis, which demonstrated that the solid substance contained calcium carbide, and the gas, acetylene. Subsequently, on the 16th of September, 1892, a specimen was sent to Lord Kelvin, in Glasgow, with a letter which has since become a famous document in scientific history, since it secures to Willson the honor of being the first to prepare calcium carbide on a scale large enough to promise commercial results.

Thus, calcium carbide is no new thing, since it has been known for the greater part of a century, but the possibility of its manufacture in quantity dates only from these experiments at Spray, in 1892.

To be sure, the French chemist Moissan and others were occupying themselves with similar investigations during the same time, and Moissan in particular has communicated to the world the results of a thorough study of this and other carbides. True it is, that there has been a legitimate controversy between the supporters of the French savant and the friends of the American experimenter, with regard to the right of priority of discovery, with the result, after the question has been thoroughly investigated, that scientific men do not fail to agree that to Willson belongs the honor. The German government has acknowledged it in a most substantial way, by annulling the German patents first granted to Bullier. Thus, through the incidental study of a by-product, and as the result of a chance accident, was the industrial possibility of calcium carbide discovered and made known to the world. To-day calcium carbide and acetylene are familiar terms upon the lips of all the people of every civilized country of the globe. Experiments were continued during the years 1892-93, which demonstrated the practicability of acetylene as a means of illumination and led to the establishment of the Willson Aluminium Company at Spray in the spring of 1895, when the first factory in the world for the manufacture of calcium carbide was installed.

LITERATURE

Previous to the last few years, the literature of our subject would naturally not be looked for outside of the most remote scientific prints, those journals which are made for the scientific classes only and read by them alone. But within a dozen years there has sprung up a voluminous popular and semi-popular literature on this subject. For those who devote particular attention to acetylene, there are special acetylene periodicals; two are regularly published in Germany, two in France, one in England and one in the United States, all journals of respectable size, responsible editorship and regular issue. Of our own American Acetylene Journal the writer can do no better than quote the following editorial comment, which recently appeared in the London Gas World: "The Acetylene Journal (published at 324 Dearborn Street, Chicago, Ill., U. S. A.) is far and away the most interesting of the acetylene periodicals that reach us from time to time. The October issue, recently to hand, is highly creditable to the publishers, both as to matter and manner. Good accounts are given in it of the progress of the acetylene industry in the States and in other parts of the world." Its genial and efficient editor and publisher, Mr. Elias A. Long, is always keenly alert to the best interests of his beloved profession, and no monthly journal, of from thirty-two to forty-eight pages, at fifty cents per annum, could possibly furnish more readable news and sound information on any semi-technical subject.

Moreover, volumes have been written, monographs and trade annuals issued, lectures delivered and published, while innumerable articles in the general press, as well as in the popular magazines, and consular reports, have spread the knowledge of acetylene and calcium carbide to the remotest corners of the globe.

Further, the patent literature of all countries has accumulated bulky volumes, first on the manufacture of carbide and secondly on the development of generators. Acetylene machines have presented to amateur inventors an attractive field, both because of the obvious usefulness of these generators, and because of the readiness with which new ideas as to the feeding of carbide and water, one to the other, can be devised. Thousands of people from all walks of life have planned new forms, and patents have been applied for in great numbers in all civilized countries.

For a further view of acetylene literature, the reader is referred to the bibliography of this subject, which appears at the end of this pamphlet.

THE CARBIDES IN GENERAL.

By a carbide, those who familiarize themselves with the concise language of science understand nothing more nor less than a compound or substance composed of charcoal and a metal. But these two parts are so blended into each other, chemically combined, that the characteristics of both are entirely lost in the properties of the product. The *carbonates* are an entirely different class of compounds, containing oxygen in addition to carbon and a metal, possessed of entirely different physical and chemical properties, and are in no way to be confounded with the *carbides*. Limestone and dolomite are natural illustrations of carbonates.

So far as we know, there may be carbides of all the metals, and many of these are known. Thus potassium carbide and sodium carbide have been studied and the results published; we can read, if we wish, about the carbides of barium and of aluminium, of magnesium, copper, silver, etc. Carbides of iron, or iron carbides, play an important role in the metallurgy of iron and steel, but the two carbides which are of leading commercial importance are those of silicon and of calcium. The former constitutes carborundum, a most useful and highly valued abrasive material, while the latter is of importance for only a few valuable properties which it possesses in itself, but chiefly because from it can easily be made the valuable illuminant acetylene. The carbides, as a class, are solid substances, not easily melted, uninflammable, hard, unattractive in appearance (carborundum is an exception to this), and about two and a half times as heavy as water. They possess in common no valuable property except that when brought into contact with water, certain of them, but not all, have the power of entering into chemical combination with the water, after the manner in which lime slakes with water, but with copious liberation of bubbles of gas. Of those which do this the one of greatest practical importance, and the cheapest, is the carbide of calcium, or we may use interchangeably at pleasure, with equal propriety, the slightly different name, calcium carbide.

Carbides do not exist in nature at the surface of the earth, though their existence in the earth's interior may possibly have been a matter of great geological importance, giving origin, as some think, to the natural supplies of petroleum, and to natural gas.

CALCIUM CARBIDE IN PARTICULAR.

This compound and its peculiarities should here be described in considerable detail. Deferring the method of its preparation to a later section, it may here be mentioned that calcium carbide is a hard, opaque, crystaline substance known to the chemist as Ca C_2, which, not to be alarmed at the suggestion of a chemical formula, is merely a shorthand expression for one unit of mass of calcium, the metal which exists in lime, combined with two mass units of carbon, the element of which charcoal is composed. Calcium carbide is therefore nothing more nor less than a compound which contains calcium and carbon only. The calcium in it is about sixty-two and a half per cent. and the carbon is thirty-seven and a half per cent. of the weight of the carbide. We all know that the common ore of iron, which is so familiar, is composed of iron and oxygen, and yet the compound is far different from the iron which it contains and which we know can be smelted out of it, also very different from oxygen, the air which we breathe all the time; we are quite accustomed to realize that the iron is there in the red ore, and the oxygen is there too, though in a substance having properties quite different from either. It should therefore be no difficult matter for us to understand that calcium, a metal, and carbon, a substance like coal, are both contained in this material, calcium carbide, though the compound is so different from either constituent.

Calcium carbide is generally of a dark brown or black color, sometimes described as dark gray, and sometimes as a bluish black, and often possessing a reddish tinge. If broken while hot, the fracture exhibits considerable iridescence. It is brittle and is more or less crystalline in structure. Its specific gravity is 2.22 to 2.26; it endures heating to redness without melting or suffering other change, though it softens and fuses under electric heat; it will not burn, except when highly heated in oxygen; it looks like a mass of stone. A cubic foot of the crushed material weighs one hundred and thirty-six pounds and in weight, color and many physical characteristics it is frequently compared with granite.

Calcium carbide may be preserved any length of time if kept sealed from air, but the ordinary moisture of the atmosphere gradually slakes it and it becomes changed on long standing into slaked lime, not essentially different from other air-slaked lime. This slaking power gives it a name, which is sometimes heard, Acetylene Lime. It always possesses a penetrating odor, which, however, is not due to the carbide itself, but to the fact that it is constantly affected by moisture, yielding minute quantities of acetylene, which accounts for the odor rather than the carbide itself. It is not affected by solvents, such as carbon bisulphide, petroleum, chloroform, ether or benzol, but if allowed to come into contact with water, or any mixture containing water, an immediate and vigorous decomposition takes place, evolving liberal quantities of gas; it is unaffected by shock, concussion or age.

Carbide is a safe substance to store or transport under proper conditions. Transportation lines accord it the same classification rates as various other forms of hardware. It cannot explode, take fire, or otherwise do harm, being similar to lime in this respect also. Since even the slow action of the moisture ordinarily present in the air will in time render carbide entirely useless, it becomes absolutely necessary, in order to securely preserve it, to pack it in perfectly tight, closely sealed containers, generally drums or cans. Granted protection from water, no substance can be safer or less likely to cause trouble when stored or conveyed from place to place. Lime becomes very hot when acted upon by water; so also does carbide, and the only essential difference is that carbide, when it slakes with water, yields a combustible gas. But this is the very fact which makes it interesting and valuable. Carbide decomposes with water in accordance with the following chemical equation:

$$\underset{\text{Carbide}}{CaC_2} \quad \underset{\text{and}}{+} \quad \underset{\text{water}}{2H_2O} \quad \underset{\text{yield}}{=} \quad \underset{\text{slaked lime}}{Ca(OH)_2} \quad \underset{\text{and}}{+} \quad \underset{\text{acetylene.}}{C_2H_2}$$

An easy calculation shows that sixty-four unit-weights of carbide require thirty-six of water, producing also seventy-four unit-weights of slaked lime and twenty-six of gas. But as gas is more conveniently and commonly measured than weighed, a better statement of the same fact would be: A pound of absolutely pure carbide yields five and one-half cubic feet of acetylene. Yet absolute chemical purity is not a practical commercial possibility, and it is natural to suppose that commercial products of different manufacturers will vary in purity. Calcium carbide is no exception. Special reasons for this will be observed later, but in practice good carbide may generally be expected to produce never less than four and a half cubic feet, measured at ordinary temperature, and rarely more than five and one-fourth cubic feet per pound of carbide; it may ordinarily be calculated at five. It is important to remember that the exact proportional weights of water and carbide must always enter into combination, i. e., sixty-four of actual carbide to thirty-four of water, be it ounces, pounds or tons; also that the same proportional weight of acetylene must always result. Any apparent exception can only be due to impure carbide, loss of gas, or failure to secure complete consumption of material.

The evolved gas does not take fire of itself but is easily lighted. Let anyone take a tumbler of water and drop into it a piece of carbide, the size of a pea, then let him light the bubbles of gas as they rise to the surface. In no way is it easier to secure a practical knowledge of the characteristic properties of this wonderful stone.

The gas burns with a bright light, not brilliant, for it needs control to bring out this property, and much smoke. In fact, soot in solid flakes may be seen floating through the air when the experiment is performed in this rough way. While carbide, as stated above, is not inflammable, nor possessed of any explosive property, such as is sometimes erroneously attributed to it, it might, by finding access to water in a closed vessel, create such a pressure as to burst the vessel, and it is not impossible that the heat generated might cause an ignition of the gas with resultant disaster. But any such condition as water and carbide gaining access to each other in a confined space is most improbable, could only occur by intention, and may be dismissed from the list of possible accidents.

When water is sprinkled or sprayed upon carbide, in not excessive quantities, the resultant slaked lime is left in a perfectly dry and dusty condition, occupying considerably more space than the original carbide. When more than enough water is employed, the residuum will, of course, be wet, either pasty or thinner if large excess of water is employed.

THE MANUFACTURE OF CARBIDE.

In the manufacture of carbide it is well to remember that the materials employed are among the most refractory ones which we know. Lime is so infusible that it is frequently employed for the material of crucibles in which the highest melting metals are fused, and for the pencils in the so-called calcium light, because it is capable of withstanding extremely high temperatures. Carbon is the material employed in the manufacture of arc lights and other electrical appliances as well, partly for the same reason. Yet in the manufacture of carbide these two most refractory substances must be forced into combination with each other.

It is the excessively high temperature attainable in the modern electric furnace, five to seven thousand degrees Fahrenheit, which alone accomplishes the combination of these elements to form calcium carbide, and it is not necessary to attribute it to the action of any electric force or process. The electric "arc" being formed in the furnace, a thoroughly incorporated mixture of ground coke and lime in right proportion is introduced. The change which takes place is:

CaO	$+$	$3C$	$=$	CaC_2	$+$	CO.
Quick lime	and	coke	yield	calcium carbide	and	carbon monoxide.

which means that fifty-six pounds of lime and thirty-six of coke make sixty-four pounds of carbide and liberate twenty-eight of carbon monoxide, a gas which escapes or is burned at the mouth of the furnace. Thus for each pound of carbide made, is consumed a pound and a half or more of a mixture which is something like seven-twelfths lime with five-twelfths coke. Granted pure material, there is formed an ingot of very pure carbide, surrounded by a crust of less pure product because partially unconverted. In breaking up, packing and shipping the carbide, this poorer crust is rejected. At first impure materials were employed for the manufacture of calcium carbide, but this resulted in an inferior grade of carbide, which in turn yielded an impure gas, so that at the present time it is everywhere recognized as essential that only first-class materials should be used. It is customary to employ lime which is ninety-nine per cent. pure, and coke of low ash. Both must contain as little sulphur and phosphorus as possible.

The Union Carbide Company, of New York and Chicago, is at present practically the sole manufacturer of calcium carbide in the United States, and more will be said of their manufacture and the distribution of their product in a later section.

Of impurities in carbide, and consequently in acetylene, discussion will be found under a special topic.

WHAT IS A GAS?

For a proper understanding of acetylene, though no more necessary than for an intelligent conception of many of the incidental materials which influence our every-day life, one ought to possess a clean-cut conception as to the meaning of the word *gas*. In practice many ordinary substances are capable of assuming, under different conditions of temperature, etc., three different physical forms.

Water is the most familiar and universal illustration. We use it in its solid form as hard, transparent ice, which everybody knows. When heated it melts, and exists as the abundant liquid which is so common everywhere. Heated still further, above two hundred and twelve degrees Fahrenheit, it boils and becomes converted into invisible steam, which, though a greatly attenuated form of matter, is still the same substance as the ice or the water from which it was produced. When we reverse the changes and cool the substance, we find that at two hundred and twelve (boiling point or liquefying point) steam becomes water, and at a still lower temperature, thirty-two degrees Fahrenheit (melting point or freezing point), it solidifies again into ice.

Now water is only one of a thousand substances which undergo just such changes, though each separate substance usually possesses its own characteristic melting point and boiling point. Thus, below its melting point, every substance is a solid; from its melting point to its boiling point it is a liquid; above its boiling point it is a *gas*. It is with this latter form of matter that we now have chiefly to do. Aside from steam, other familiar gases are air, hydrogen, illuminating gas, etc., but most of these are mixtures of two or more individual gaseous substances. Thus air is a mixture of oxygen and nitrogen; illuminating gas is a very complex mixture of numerous gases, etc. All gases tend to diffuse with each other when opportunity allows, and all tend to occupy all available space. Gases are extremely attenuated as shown by the fact that a cubic inch of water makes a cubic foot of steam. The minute particles which compose a gas must therefore be much farther apart than in liquids or solids. Physicists believe that these minute particles of matter, or molecules, are in constant motion, and possessed of attractive power which holds them firmly together so long as they are in the solid form; that heat accelerates their motion until presently the molecules move rapidly enough to overcome the attractive force to such an extent that the solid becomes liquid; that in the liquid form the molecules still stick together, so to speak, to such an extent that the substance flows, seeks its lowest level, etc., whereas, when further heat increases their motion still more, the molecules fly away from each other and the gaseous condition is produced.

Steam, air and almost all gases are colorless and invisible; an incandescent lamp bulb from which the air has been exhausted *looks* just the same as if it were filled with air. But notwithstanding that they are hidden from our vision, air and steam and acetylene are just as real as any other kinds of matter with which we have to do. No one will deny the reality of the material which fills a distended automobile tire, nor of that which operates within the cylinder of a locomotive. Manifestly, however, the gaseous state is one of extreme expansion and gaseous volumes consist mainly of intermolecular space; in fact, it would not be a misstatement to say that a cubic foot of acetylene consists of much more than niney-nine per cent. of unoccupied space, and far less than one per cent. of space totally filled with matter. This leads to the most characteristic property of gases, which distinguishes them from liquids and solids, namely, their almost perfect elasticity. Owing to this property, gases can be condensed by pressure, and the volume yields exactly in proportion to the pressure exerted upon them. Press twice as hard upon any quantity of gas as you did before and you halve its volume; reduce the pressure to a half and you double the volume. Sufficient pressure compels most gases to assume the liquid condition.

In like manner, all gases expand when heated and contract when cooled, these changes also taking place under a precise and simple law, so that it may properly be stated that the volume of a gas is directly proportional to its absolute temperature.

It follows from the above facts, since gases are perfectly elastic bodies, whereas liquids are almost entirely devoid of elasticity, that though a cubic foot of water is a fairly definite quantity, being always one thousand ounces at ordinary temperature and varying only a few per cent. in weight from thirty-two to two hundred and twelve degrees, nevertheless a cubic foot of acetylene, or any other gas, is not in any sense a definite quantity, unless it be expressly stated at what temperature and pressure the gas is to be measured.

Gases vary greatly in weight, according to their composition. Hydrogen, the lightest, is one-fourteenth as heavy as air; steam weighs five-eighths as much as air. Many are lighter and many are heavier; a balloon to float in air must necessarily be inflated with a lighter gas.

Many gases are inflammable; many are not. Those which are capable of burning do so with varying amounts and colors of flame, which may or may not be smoky, and with the production of greater or less quantities of heat and light. No two gases are alike in these respects, but every individual gas behaves always in the same way, and every gas which can be burned at all requires its definite quantity of oxygen to maintain its combustion. This oxygen is commonly derived from the air and without it no gas or other fuel whatever can be burned. If the gas burns quietly and slowly, as in an ordinary burner, the oxygen is regularly supplied from the surrounding atmosphere and the process continues without interruption or violence. But if a quantity of an inflammable gas is first mixed, intentionally or by accident, with a sufficient volume of air to accomplish its combustion, and then ignited, an explosion results, because the total quantity burns in a moment, and suddenly demands and occupies so much space that objects in the neighborhood are compelled to give way.

An explosive mixture of gases is therefore such a mixture of a combustible gas, with air or oxygen, as contains at the outset the right proportions of the two to satisfy each other.

The products of combustion, since no matter is ever annihilated, and since combustion merely produces new forms, must also be reckoned with. These are usually gaseous also and many possess objectionable or even poisonous character, according to circumstances. The coal miner fears not only the dreaded fire-damp, which by mixing with air and becoming ignited may demonstrate the terrific violence of its explosive character, but he has reason also to shun the deadly after-damp, which may so vitiate the atmosphere that he may escape the explosion only to fall a victim to the stifling action of the gases which follow in its wake.

GENERAL PROPERTIES OF ACETYLENE.

Acetylene, whether prepared from carbide or in any other way, is known to the chemist as C_2 H_2, meaning that it has the constant composition of twenty-four parts by weight of carbon and two of hydrogen, or ninety-two and three-tenths per cent. carbon and seven and seven-tenths per cent. hydrogen. Being composed of these two elements only, it belongs to the class of compounds known as hydrocarbons, represented in nature by natural gas, petroleum, etc. All hydrocarbons are combustible, forming as products of combustion, carbon dioxide and water. There are hundreds of hydrocarbons known, of greater or less importance, and more or less thoroughly studied, and there are many thousands of compounds containing carbon united with elements other than hydrogen, but among them all there is no gas containing a higher percentage weight of carbon than acetylene. It is, therefore, proper to look upon acetylene as essentially gaseous carbon. Carbon, known in so many forms as a fuel for combustion, with a view either to producing heat or furnishing light, a constituent of all fuels from whatever source and wherever employed, is in no place to be had more closely approaching the condition of pure carbon in the gaseous state than in acetylene. Pure carbon, one hundred per cent., as well as in the various commercial forms, anthracite, coke, charcoal, etc., does not even melt under such heat as can be attained, much less pass into the form of vapor or gas. Hence it is interesting to note that acetylene, containing only seven and seven-tenths per cent. of material other than carbon, is the nearest approach to gaseous carbon at common temperatures with which we are acquainted.

Acetylene is a colorless and tasteless gas, possessed of a peculiarly penetrating and persistent odor. In describing the gas as it is most frequently known, this characteristic odor should not be ignored. Its presence in the air to the extent of only one part in ten thousand is distinctly perceptible. It should be observed, however, that in burning properly at the jet there is absolutely no odor perceptible, *and if an odor of acetylene is detected about an apparatus in operation,* it is *certainly* because of the *leakage of the gas,* through faulty piping, an open cock, or otherwise.

Acetylene dissolves in water to a considerable extent. This is usually stated as thirteen volumes of acetylene in twelve volumes of water. This statement is for ordinary temperature, approximately correct, and it is near enough for practical purposes to state that a gallon of water may dissolve a gallon of gas. It is more soluble in certain other liquids such as ether, alcohol, petroleum, turpentine, and especially in acetone; we shall have occasion to refer to this last solvent in a later chapter. Since it is most frequently generated in contact with water which, by the very nature of the circumstance, contains more or less lime, its exact solubility in such water, at different temperatures, is of interest here. The following figures show the solubility of acetylene in clear saturated lime water, and in milk of lime containing ten parts by weight of quicklime in one hundred parts of water, the figures representing the number of volumes of acetylene dissolved in one hundred volumes of the fluid:

Degrees Fahrenheit.	Lime water.	Milk of lime.
32	146.2	152.6
41	138.5	.....
68	122.8	134.8
112	43.9	62.6
194	6.2	9.2

The full measure of the solvent power of water and other liquids for acetylene is not reached unless the gas and the liquid are given ample opportunity to come into free contact with each other. Thus if the gas be kept within a receptacle over water which is at rest, the surface layer of water to a slight depth becomes saturated with the gas, protects the lower layers of water from becoming so, and the further penetration of the gas into the water is retarded or even stopped. But if the water is agitated in any way with the gas, or the gas allowed to bubble up through the water, the water easily takes up its full allowance of gas, and under such circumstances the withdrawal of a quart of water involves also the removal of the contained quart of gas. This idea of the solubility of a gaseous substance in water is a peculiarly difficult conception to grasp, but if we may just accept it as a fact, and consider it parallel with the solution of salt or sugar in water, we shall be on the right track in this respect. Acetylene is lighter than air in the proportion of ninety-one to one hundred; its specific gravity is therefore said to be 0.91. A balloon filled with it will consequently float in air. Yet it is much heavier than coal gas, whose average specific gravity is 0.43, or natural gas, 0.56.

Acetylene burns in the air with a brilliant but smoky flame. When its illuminating power is properly brought out, as it may be by the use of a suitable burner, it yields a light of greater brilliancy than that furnished by any other gas.

All hydrocarbons burn in air with production of carbon dioxide and water, but the quantity of combustion products, compared with the volume of gas burned, varies with the character of the hydrocarbon used.

The combustion equation,

$2\,C_2H_2$	$+$	$5\,O_2$	$=$	$4\,CO_2$	$+$	$2\,H_2O$
Acetylene	and	oxygen	yield	carbon dioxide	and	water.

means that two volumes of acetylene burned require five volumes of oxygen and yield four volumes of carbon dioxide and two of water vapor. Better stated, one cubic foot of gas burned deprives the surrounding air of two and one-half cubic feet of oxygen, throwing into it two cubic feet of carbon dioxide and one of water vapor. Here, of course, is meant the complete combustion as it takes place in a perfect burner, and space will not be consumed in considering imperfect or retarded or partial combustions which are possible, but are aside from the line of importance.

All gases which burn in air will, when mixed with air previous to ignition, produce more or less violent explosions, if fired. Common illuminating gas, issuing quietly from the jet where it has been lighted, burns regularly and safely with no suggestion of explosion. But allow the gas to become mixed with air as by accidental leakage into a cellar or closet or other confined and unventilated place, then bring about its ignition by applying a flame as when a match is lighted in the vicinity, or when one goes into the dangerous place with a lighted lamp or candle, and an explosion is sure to result which is more violent as the mixture more nearly approaches the proportions necessary for combustion. This is a perfectly general fact, as true of one combustible gas or vapor as another. Creative Power endowed all ignitible gases with this property, and it is as sure to be true as water is to flow, be it in the coal mine, at the gas works, along a leaky city main or in the bowl of an overheated oil lamp.

Of course, acetylene is no exception. From the statements just presented it is easily calculated that one measure of acetylene and twelve and a half of air are required for complete combustion; this is therefore the proportion for the most perfect explosion, i. e., mixtures of acetylene and air which contain about seven and one-half per cent. of the gas are expected to explode with the most terrific violence. But let it not be supposed that a seven and a half per cent. mixture is the only one which can explode with violence; all mixtures from three per cent. to thirty per cent. of acetylene in air are capable of exploding with greater or less force, according to conditions, if ignited.

It may not be out of place here to note that a great deal of undigested misinformation has been given to the public with regard to this point. Errors may have crept into the statements regarding the explosive limits of acetylene air mixtures, and been copied from place to place. The author of this paper, in its former edition, gave these limits very differently, but he desires now to record his view that it is not possible to secure an explosion, in any true sense, by presenting a flame to a mixture which contains more than thirty per cent. of acetylene. Numerous experiments have been made in his laboratory, by competent chemists under his eye, and at no time has an explosion been secured, though bare flame ignition was attempted in a great variety of ways, when the acetylene was present in larger proportion than that mentioned. He thinks it therefore a proper correction to make on the data which have often been published before.

The ignition point of acetylene is lower than that of coal gas, being about nine hundred degrees Fahrenheit, as against eleven hundred degrees. This is easily exhibited by allowing the gas from an acetylene burner to blow on to a cigar burning in the usual way without flame, when the gas will be ignited. Common illuminating gas cannot be lighted in this way

Under comparatively light pressure, twenty-six atmospheres, at thirty-two degrees Fahrenheit, or about forty atmospheres at ordinary temperature, a measure of compression easily reached with modern pumping appliances, acetylene gas becomes a liquid. This liquid is said to be the lightest liquid known, being only four-tenths as heavy as water. It is possessed also of an expansibility greater than that of any other known liquid. Acetylene may further be had in a solid form, resembling snow, and this is called acetylene snow, bearing as it does to liquid acetylene and gaseous acetylene the same relation that ordinary snow bears to water and steam respectively.

Acetylene in the gaseous form, at ordinary pressure, is not explosive in any sense except as referred to above, when mixtures of the gas with air become ignited. So important is the correct understanding of the exact facts in the matter that, even at the expense of some repetition, it should be clearly understood that acetylene, not compressed, cannot be exploded, detonated or otherwise set off in any manner whatever that could warrant its being called explosive. A lighted candle plunged into it will be extinguished, setting fire to it, of course, where it comes in contact with air, for air is necessary for its combustion; but no combustion or explosion can take place in the midst of a body of the gas. Even a fulminate cap exploded in the gas will not propagate a wave of explosion to any perceptible extent whatever, nor will an electric spark in an atmosphere of the gas cause any injurious or dangerous result whatever, except such as would result from lighting any other gas in the same manner.

But the liquefied gas, liquid acetylene, is quite another matter. Under certain conditions, not necessary here to analyze, liquid acetylene can be exploded, and for that reason liquid acetylene may be set down as a dangerous substance. Even this preparation, when our knowledge of it becomes more perfect, may perhaps be so controlled as to render its restricted employment safe, but for the present a wise rule is to allow no liquid acetylene to be handled under circumstances where its explosion might do harm. In fact, it would be wise to go still further and avoid the use of acetylene under any pressure which might by any possibility approach that which is necessary to liquefy the gas. The liquefying pressure has been mentioned as twenty-six atmospheres at thirty-two degrees Fahrenheit; this means nearly four hundred pounds per square inch at freezing temperature and about seven hundred pounds at seventy degrees. This pressure increases with temperature up to ninety-eight degrees Fahrenheit, which is its *critical temperature,* above which point no pressure will suffice to hold acetylene in a liquid state.

No one could, by any reasonable possibility, accidentally liquefy any acetylene, and so the safeguard is a perfectly easy one to apply, and in all appliances in which acetylene is prepared and used by the public, the likelihood of creating a dangerous compression is reduced to nothing.

It is still further true that acetylene, at some pressures, greater than normal, but not approaching that necessary to liquefy, has under most favorable conditions been found to explode; but it may be said with perfect confidence that this gas has been studied a very great deal and its properties are as well known as could be wished, yet under no circumstances has anyone ever secured an explosion in it, if subjected to less than one atmosphere of pressure. And as the limit of one atmosphere, fifteen pounds to the square inch, is a perfectly easy one to observe, and as in the usual domestic apparatus there should never be a pressure of more than a few ounces, the conditions of safety along this line need by no means be overreached.

There are, it is true, applications of acetylene of a very important character where the gas requires to be used under pressure, but these do not pertain to the ordinary household use of acetylene as an illuminant, and wherever they are applied abundant precautions may be taken to render them safe. In fact, these applications do not extend in any measure to the uses which the general public makes of acetylene, and when they are practiced it need be only by trained specialists and skilled workmen who thoroughly understand the conditions under which they work. More will be said about these applications under such topics as Dissolved Acetylene, Oxy-Acetylene Blowpipe, etc.

Though not exploded by application of high heat, acetylene is injured by such treatment. More properly speaking, it is partly converted, by high heat, into other compounds, thus lessening the actual quantity of the gas, wasting it, and polluting the rest by the introduction of substances which do not belong there. Among such compounds into which acetylene, $C_2 H_2$, may be partially converted by heat are benzol, $C_6 H_6$, and styrolene, $C_8 H_8$. These and others of their like remain in part with the gas, causing it to burn with a persistent smoky flame, which cannot be corrected, and producing carbonization of burners, and in part deposit a solid tarry residue at the point where the heating takes place. In cases where the gas is generated without avoiding undue rise of temperature objectionable results as above are certain to be secured.

Compounds of acetylene with copper and silver have long been known and they are violently explosive substances. They are easily made in the laboratory if one takes the trouble to make the experiment, and it has been supposed that the copper compound might be formed by action of the gas on copper or brass containers, pipes, parts of apparatus, fixtures, etc. But careful investigations, during which acetylene has been allowed long contact with copper and brass, under more varied and trying conditions than would ever be likely to exist in practice, have been repeatedly and laboriously carried out, and the idea that such conditions can result in the accidental formation of dangerous substances has been entirely set aside as disproved.

SPECIFIC PROPERTIES OF ACETYLENE.

Acetylene an Endothermic Compound. A great deal has been said by those who feel that they can afford to use scientific terms, about acetylene being an endothermic substance. This bugbear may as well be disposed of once for all, and we shall find that the name is a great deal worse than the condition which it describes, as regards acetylene at least. There are certain compounds during the formation of which heat has been absorbed, instead of evolved, as is the usual rule. Such compounds are scientifically described as endothermic, and some of them, because of their power of developing much heat when they decompose, are very explosive substances. Acetylene is endothermic, and if pressure be exerted upon it, forcing its particles nearer together, and compelling it to occupy less than its accustomed normal amount of space, it develops the character of an explosive. The only trouble with the use of the word resides in the danger of the inaccurate assumption that an endothermic substance is of necessity also an explosive one. Acetylene at ordinary pressure is not an explosive, and it does not become so, as stated above, until at least one atmosphere of pressure, and probably two, in addition to normal, have been added; these are conditions which never occur in connection with a domestic generator.

Physiological Effect of Acetylene. Statements have been made to the effect that acetylene is poisonous. These statements are in a general way untrue, though it could not properly be stated that they are wholly without foundation. Reliable investigators, Berthelot, Rosemann, Gréhant, and others, whose conclusions are not to be questioned, have independently of each other carried on researches not necessary to detail here but easily referred to when desirable, looking to the establishment of accurate information on this point, and have shown that while it is slightly poisonous, it is less so than coal gas and vastly less poisonous than water gas.

There is a gas called carbon monoxide, well known to exert a most deleterious action on the blood whenever it is inhaled; this gas is responsible for the poisonous action of furnace and stove gas when, through faulty drafts, household heaters throw out noxious vapors over night and asphyxiate the householders in their beds. It is the deadly gas which sometimes overcomes persons who breathe the fumes of charcoal burning in confined rooms. It issues from blast furnaces, from kilns of various sorts, and workmen about metallurgical operations know full well that it is to be avoided. It is insidious and one never knows he is breathing it till the poisonous symptoms commence, for it is without odor, devoid of taste and color, and it kills! A prominent authority states on this topic, "It is certainly fatal to breathe air containing one per cent. of carbon monoxide, and it is said that so little as one-twentieth of one per cent. may prove fatal." The writer does not wish to decry the use of water gas in our American towns; on the contrary, he believes that sufficient safeguards are usually exercised against the dangers and that our city gas supplies, as a rule, call for no serious criticism on this score, although city gas in America runs as high as twenty per cent. of carbon monoxide, and in some localities accidental deaths from inhaling coal gas are of too frequent occurrence.

When we reflect that "blowing out the gas" is a foolhardy act that everybody knows to be dangerous, and that every great daily paper has frequent occasion to report cases of burners left open with suicidal intent, is there not great satisfaction in the knowledge on the part of a consumer of acetylene that, with equal lighting power, he is using an illuminant so free from deleterious quality that leaving a jet open in a living room for a week involves no peril to human life?

Nor does it involve undue enthusiasm for acetylene if its adherents proclaim that this quality brings it virtually into the class of life savers! For, just in so far as it replaces poisonous illuminants, it is a direct means of averting death.

In one of Gréhant's experiments upon dogs a mixture of twenty per cent. of acetylene with air inhaled for thirty-five minutes did not seem to trouble the animal. A dog breathing a similar mixture of illuminating gas containing only one per cent. carbon monoxide quickly showed convulsive movements and died after ten minutes. The way in which carbon monoxide, the poisonous constituent of coal gas and water gas acts, is by combining with the hæmaglobin of the blood, producing death by asphyxiation, whereas reliable investigation has proved that such acetylene as enters the system under similar circumstances is freely given off again without injury.

Acetylene is irrespirable, so indeed is nitrogen or hydrogen, but it is not to be claimed that it is toxic in its action on animal life in any such sense as carbon monoxide is. A long time of inhalation and large quantities of acetylene are necessary to bring about symptoms of illness in animals.

Explosibility of Acetylene. It has already been stated that there are two ways in which an explosion can be secured with the aid of acetylene: First, when mixed with air and ignited; secondly, when, under abnormal pressures, it is set off by concussion. It is important that these two conditions should now be examined a little more closely.

1. Mixtures with Air. It has been observed under a previous topic that all mixtures from three per cent. to twenty-five or thirty per cent. of the gas with air explode when kindled. This is the force which is relied upon for the actuating power in gas engines, and acetylene should meet a broad application in this direction.

In the early days of acetylene lighting it had been suggested to mix the gas with air previous to conducting it to the burner, the object being to secure a smokeless flame. But in view of the dangerous character of such a mixture, and in view also of some accidents which occurred, this method was given up after the invention of burners, which accomplished the same purpose by effecting a mixture with air right at the point of combustion.

2. Explosions of Acetylene under Pressure. In the first years of the acetylene industry one of the fondest hopes of its promoters was that the transportation, storage and sale of the product condensed into liquid form might become an assured success. Some four hundred cubic feet of gas are required to make one cubic foot of liquefied acetylene, and when the pressure is released the whole

quantity of the gas is again secured. The scheme was an ideal one. Houses, steamboats, streets or railway cars were to be lighted by gas drawn with the aid of a reducing pressure valve from a cylinder of condensed acetylene kept at any convenient point and connected by a pipe to the usual distributing system. But explosion after explosion occurred, many of them serious and disastrous, in which the cause was always traced back to the use of liquid acetylene. Many and thorough investigations were made with the result that liquefied acetylene was shown to be a highly dangerous substance. Whether the time will come, as the enthusiasts fully believed, when liquid acetylene can be made and employed with safety, when it shall be known how to so prepare and handle this product as to eliminate the danger in its use, cannot now be said. At present liquid acetylene must be classed as a very hazardous material. Insurance companies very rightly refuse to have anything to do with it. Commercially there would be no gain, inasmuch as carbide itself is a more condensed form for shipping and storing the gas than is the liquid condition. A cubic foot of crushed carbide represents between five and six hundred feet of gas; a cubic foot of liquid acetylene represents only four hundred feet. Fortunately, also, we now have a good and sufficient substitute, which will be described in the chapter on "Dissolved Acetylene."

Combustion and Illuminating Power of Acetylene. Controlled by a suitable burner, the flame of acetylene is absolutely white and of intense brilliancy. In quality it is the nearest approach to daylight that we know. Its spectrum closely resembles that of sunlight and consequently all colors appear the same as by daylight instead of being distorted as by gas light, candle, oil or electric light. This property renders it very desirable for ordinary domestic purposes, and also adapts it admirably to photographic uses, to matching colors and accomplishing other work at night which is impossible with other illuminants, to forcing plant growth, and many similar purposes.

Light for light, acetylene illumination causes much less elevation of temperature in the atmosphere of rooms than do other forms of lighting with gas or oil; it therefore possesses a distinct advantage in this respect, and while it may be misleading to state that acetylene burns with a cool flame, it is correct to say that an acetylene burner produces much less heat than a flat flame coal gas burner which emits the same quantity of light.

Acetylene also pollutes the air less than coal gas, and we may calculate the accuracy of this statement somewhat as follows:

Recalling the combustion formula of acetylene:

$$\underset{\text{Acetylene}}{2C_2H_2} \underset{\text{and}}{+} \underset{\text{oxygen}}{5O_2} \underset{\text{yield}}{=} \underset{\text{carbon dioxide}}{4CO_2} \underset{\text{and}}{+} \underset{\text{water.}}{2H_2O}$$

it was pointed out that one meaning to be derived therefrom is that when one cubic foot of acetylene is burned, it takes from the air around the flame two and one-half feet of oxygen and throws out two feet of carbon dioxide gas. Both of these operations vitiate the air in an enclosed room, just as does the burning of any other fuel or the breathing of persons or animals. In the open air, or in perfectly ventilated apartments, this might be ignored, but under usually prevailing conditions, the injury to the atmosphere is by no means insignificant. One foot of coal gas of average composition requires one foot of oxygen and throws out three-fourths of a foot of carbon dioxide; but since only one-half of a foot of acetylene will ordinarily be burned when five feet of illuminating gas would be consumed, it is necessary, for the comparison, to multiply these last results by ten, and it will be seen that light for light a burner consumes—

Acetylene, 2½ feet oxygen, yielding 2 feet carbon dioxide.
Coal gas, 10 feet oxygen, yielding 7½ feet carbon dioxide.

Or, acetylene in producing the same candle power, impoverishes the air one-fourth as much, and pollutes it also one-fourth as much as illuminating gas. The calculations of Professor Lewes and other authorities are still more favorable to acetylene than are these figures.

"As an illuminant acetylene surpasses in lighting power and economy all other illuminants known; when burned at the rate of five cubic feet per hour it produces light equal to two hundred and fifty candles, whereas, the best illuminating gas made from coal, or water gas, rarely exceeds twenty-two candles for each five feet burned per hour. Your Philadelphia city gas is rated at from nineteen to twenty candles. Acetylene will, therefore, produce twelve and a half times more light if the same quantity be consumed, or one thousand cubic feet of acetylene gas will give you the equivalent in lighting power of twelve thousand five hundred cubic feet of your city gas; it has, therefore, twelve and a half times the value." These were the words of Willson and Suckert, quoted from their paper before the Franklin Institute, March 17, 1895, and though that was one of the earliest recorded statements from persons who can be regarded as authoritative, later research has not had occasion to greatly modify their figure. The relative light value of acetylene and illuminating gas is variously stated, often in exaggerated terms, especially in the expressions of those whose chief motive is to sell a machine. But thirteen is a frequent figure now used by good authorities, and the writer considers that, for a conservative and safe figure on which to base any calculations that this paper may call for, one cannot do better than use the ratio set down by the authorities just mentioned, namely, that foot for foot, acetylene has twelve and a half times the illuminating power of city gas. In all such comparisons it is well to remember that acetylene is a compound, a single or individual substance, of invariable composition the world around, wherever it may exist or be prepared, except for occasional or accidental admixture of other material, whereas coal gas is not a compound, but a mixture, of variable proportions, different for every different gas works, though of course possessing an average composition which can be used as a basis of calculation, while the product of individual plants may differ materially from such average.

The comparison of the illuminating power of our gas with that of other well known compounds, as made by Lewes, is as follows, all based on five cubic feet per hour of consumption:

1. Methane, 5.2 candles.
2. Ethane, 35.7 candles.
3. Propane, 56.7 candles.
4. Ethylene, 70.0 candles.
5. Butylene, 123.0 candles.
6. Acetylene, 240.0 candles.

This table is of interest because substances (1) to (5) are all constituents of coal gas, in varying proportions. But coal gas, as regards its luminous constituents, consists chiefly of methane, mixed with but little of the other gases in the above table, which possess a higher illuminating effect, and mixed at the same time with large percentages of gaseous matter possessing no luminosity whatever. Hence the low average lighting power of coal gas which for Philadelphia city gas was placed, above, at nineteen to twenty candles for five feet per hour. With the modern improved burners for acetylene, the best half-foot burners should develop twenty-five candle power as against the five-foot coal gas burner developing twenty candles. Thus the ratio of twelve and one-half to one is maintained. And though this may be variously stated by interested parties, now high, now low, it is safe to accept, as an entirely unprejudiced basis of calculation, that the highest effects which can be produced from acetylene and coal gas respectively, without the aid of artificial air supply other than that created by the flames themselves, are as twelve and one-half to one.

DANGER.

The possible danger to arise from the use of any commodity, whether a foodstuff, an illuminant or any other article of necessity or convenience, should be thoroughly known before any steps are taken.

It seems to the writer that all the dangers in the use of acetylene must classify under one of the following heads:

(a) Danger which may arise from the poisonous action of acetylene in case it should escape through leaks or open burners into the air.

(b) Danger which may arise from breathing air vitiated by the combustion of acetylene in it.

(c) Danger which may arise from explosion.

(d) Danger which may arise from fire.

A few years ago, when the commercial production of carbide was a novelty, most exaggerated accounts of its dangerous character were published in the newspapers and passed from mouth to mouth. Now, every child knows that carbide is a perfectly safe commodity, requiring only to be kept tightly closed and away from water. Transportation companies make no more objection to accepting a drum of carbide for shipment than they do to a keg of nails.

A fire in a carbide storage warehouse at Chattanooga, starting from an overturned lantern, destroyed various other material stored on the same floor and practically burned down the building, but did no damage whatever to the stock of carbide. Again, during the tremendous floods at Kansas City in 1903, a warehouse containing between thirty and forty tons of carbide was completely submerged and in the midst of the swiftest current for eight days. After the water had receded, the carbide was recovered, practically without loss.

Calcium carbide may now be shipped, on passenger vessels, to any part of the world, and all restrictions upon such shipments which formerly existed before its properties were fully understood have been rescinded.

In like manner, whether fortunately or unfortunately, about everyone who has heard of acetylene has heard also of its supposed dangerous character. Some would have it dangerously poisonous, others dangerously explosive. No doubt there were some impurities in the earlier specimens of carbide which were placed upon the market, and these caused impurities in the gas, which, whether toxic or otherwise, dangerous in any way or not, modified the character of the first few thousand feet of acetylene used. But, to-day, when we may be entirely confident of the commercial purity of our carbide, this question is not serious. The impurities of acetylene will be discussed in a later topic, and now that the first outbursts of the alarmists have long since died away, we may impartially, seriously and quite thoroughly analyze and inquire into all the possible sources of danger. Let us consider them in the order above suggested.

(a) Danger from Toxic Action. It has been shown that acetylene is far less poisonous than the usual illuminating gas sold in our cities. While it might occur to a wag to observe here that it matters little what becomes of the man who intentionally, or otherwise, "blows out the gas," it is still to be said that far less gas can pass through an ordinary acetylene burner, one-half foot, than would pass through an ordinary coal gas burner, five-foot size, assuming that both are accidentally left fully turned on and gas escaping at full head. So that if the "patient" should quietly breathe the resultant polluted air, allowing, for the sake of argument, that the two gases are of equally poisonous character, which is not true, it would take him ten times as long to get a troublesome dose of acetylene as of coal gas. But it is highly improbable that he would quietly breathe the gas long enough to do any harm whatever. Acetylene makes its presence known by its peculiar but persistent odor, which in most cases would be a sufficient safeguard. Again, its effects are irritant and annoying rather than somnific, a second important circumstance favoring the probability of the acetylene leak being discovered and stopped. Further, no case of asphyxiation by acetylene has ever been recorded. Asphyxiation, in the sense in which we know it, as resulting from the respiration of coal gas and stove gas, is entirely unknown to the character of acetylene.

(b) Danger from Vitiated Air. The combustion of all oils or gases used for illumination, and the respiration of all animals impoverish and pollute the air in an unventilated space, rendering it "close." It has been shown that it is an inherent property of acetylene to offend less in this direction, light for light, than any other illuminating gas. Since the results of burning acetylene are exactly the same in character as when petroleum or coal gas or water gas is used, namely, robbing the air of oxygen, and charging it more or less with carbon dioxide, and since exactly the same charges may fairly be made against oil or coal gas as against acetylene, and to a greater degree, it would be manifestly unfair to acetylene to dwell longer on so remote a possibility of danger.

(c) Danger from Explosion. In case of uncondensed acetylene, or any other combustible gas whatever, two things are necessary for explosion:

1. Admixture with air.
2. Ignition of the mixture.

The gas must first escape into and mix with the air, and then be ignited or kindled. In no other way can explosion result. Suppose through ignorance or carelessness a gas jet is blown out or left open. As just now stated, half a foot will pass in an hour. Suppose it to be a foot. A room ten feet square and eight feet high, scarcely more than a good sized closet, is certainly small enough to make a fair illustration. Granted that this room is perfectly gas-tight, a highly improbable condition in itself, but let it be supposed. In just twenty-four hours enough gas will be admitted to charge the air of the room to the lowest explosive limit of three per cent. Still this mixture must find a way to become kindled before an explosion can possibly ensue. And from what was just stated of the irritating, rather than somnific effect of this gas, it is highly improbable that a person would continue to remain in a closed room long enough for danger to arise from the gas escaping out of an open burner. Acetylene can, however, never escape without revealing its presence by reason of its penetrating odor, which is indeed a blessing in disguise in such cases. In the event of cracks, defective joints, etc., it is further true that far less acetylene would escape in a given time than would be the case with coal gas, because of its greater specific gravity. Distressing in truth are the accounts of accidents caused by the ignition of gaseous mixtures resulting from defective coal gas pipes, but the man who searches for a gas leak with a lighted candle or lamp is of the same type as the man who "did not know it was loaded." There is no remedy for this inherent difficulty, hence no way to remove the danger except by rigidly adhering to the rule that no light shall be taken where any combustible gas can, by any possibility and to any extent, be mixed with air.

Open flames should on no account be used in the vicinity of acetylene generators or gas holders; neither should a lighted cigar. This is a real danger, the writer makes no attempt to conceal or mitigate it; but it is a danger of which acetylene partakes, together with every other illuminating gas or oil whatever, share and share alike. It is further true that explosions of the character under consideration can scarcely ensue except from gross inattention, ignorance or carelessness, and relatively they are no more numerous than those arising from coal gas.

Everything considered, though prejudiced against it at the outset, and absolutely and purely disinterested at present, after much careful study and feeling the responsibility of being called upon to furnish an unbiased opinion in this matter, the writer believes that *the use of acetylene for the illumination of rural homes, provided it is generated from good carbide in a first class apparatus, and all reasonable regulations followed, is no more fraught with danger at the present day than any available method of illumination by gas, or electricity, and less so than the usual employment of petroleum.*

(d) Danger from Fire. Acetylene will indeed burn, else these pages had never been written. So does kerosene and so does illuminating gas. When, in England, a century ago, the promoters of coal gas, after many and protracted conventions, finally secured the permission to light the Houses of Parliament, the authoritative document contained the condition that pipes conveying the gas "shall be placed upon brackets at least eighteen inches from the wall." And later, in Philadelphia, in the early thirties, the people presented a "Remonstrance Against Lighting with Gas," which contained the following wisdom: "We consider it a most inexpedient, offensive and dangerous mode of lighting. We consider gas to be an article as ignitible as gunpowder, and usually as fatal in its results."

We shall hardly care to go backwards a hundred years and seek a ridiculous objection to acetylene. We shall rather look to all precautions against faulty machines and leaky piping, and not expect fuel to refuse to burn when we kindle it. Care is needed in all things, but some require more than others. In petroleum lighted homes one of the commonest accidents is caused by children overturning oil lamps. How frequently may we read of such a catastrophe in the daily press! Acetylene, now used in a million homes, has never, in this way at least, been responsible for the death of a child, a fact which leads to warm praise for acetylene in the home.

While this general subject will be before us again, in the chapter on insurance regulations, it will not be out of place to remark here that insurance experts have long since officially admitted that acetylene has proved itself less hazardous in proportion than other forms of artificial illumination in common use, such as gasoline, kerosene, electricity and candles.

IMPURITIES IN ACETYLENE.

With regard to impurities in acetylene, it may first be remarked that such impurities owe their origin, almost exclusively, to impurities in the carbide from which it is derived. Impurities in carbide therefore claim the first consideration. These may be grouped in two classes:

(a) Those which exert no action on water, and

(b) Those which, reacting with water, throw off some gaseous products of the reaction, which remain with the acetylene, rendering the latter impure.

The first class, consisting of such substances as carbon (coke not changed in the furnace), silica (sand), and a few others, since they can with water produce nothing which is added to the resultant gas, are harmless except in so far as they increase the quantity of the residue, and so may be dismissed.

The second class, which may impart to the gas small quantities of impurities, harmless or otherwise, should be looked into. Some years ago, when the carbide industry was altogether new, and when the attempt was made to put carbide on the market at something approaching the ridiculously low figure which was first predicted for it, cheap materials were employed, and a poor result secured. It was quite naturally supposed that the cheapest lime and the most worthless coal would answer the purpose. The result was that some very poor carbide was sold and some correspondingly impure acetylene was made. It was soon found that this would not do, the practice was corrected, and at present a specimen of carbide made by responsible manufacturers is rarely found, in this country at least, which can be said to be seriously impure.

Present practice does, and future practice must, insist on purity of material. Good metallurgical coke, the analysis of whose ash shows only the merest traces of sulphur and phosphorus, is easily obtainable. This with a reasonably pure lime yields carbide of great purity, and there should be no other made. In fact, the quality of carbide has greatly improved and it is reasonable to suppose that it will, in the future, maintain its present high standard. Absolutely chemically pure carbide is an industrial impossibility, but the product which is known as "commercially pure" is an entirely satisfactory article. But even as the best available carbide contains some quantities of impurity, it is well to see what results may occur in the quality of the acetylene.

Phosphorus and sulphur, mostly but not wholly dissipated by the heat of the furnace, appear to a slight extent in the acetylene in the form of minute quantities of phosphoretted hydrogen, PH_3, and sulphuretted hydrogen, H_2S. There may also be some small quantities of ammonia, NH_3, and traces of hydrogen, silicon hydride, and possibly one or two others of little significance.

An analysis of the crude acetylene evolved from commercial carbide in a carbide feed machine resulted as follows:

Acetylene	99.36 per cent.
Oxygen	00.08 per cent.
Nitrogen	00.11 per cent.
Hydrogen	00.06 per cent.
Sulphuretted hydrogen	00.17 per cent.
Phosphoretted hydrogen	00.04 per cent.

Ammonia	00.10 per cent.
Silicon hydride	00.03 per cent.
Carbon monoxide	00.01 per cent.
Methane	00.04 per cent.
Total	100.00 per cent.

This is a recent analysis, and while analyses vary somewhat, it is a typical one. Of the impurities considered, hydrogen and methane are harmless and may be dismissed. Oxygen and nitrogen in such small quantities are also of no consequence. They result from unavoidable admission of small quantities of air. Carbon monoxide has already been discussed and a trace can be found in most samples of acetylene, but its quantity is so exceedingly small as to be entirely negligible. Phosphoretted hydrogen and silicon hydride are self-inflammable and hence would kindle the gas, on exposure to air, if either could be present in considerable quantity. This is not so serious a property as might at first seem, for no kindling could occur until the gas escapes into the air, and such ignition, if it ever occurred, could scarcely do harm. Such a thing as kindling the gas within the generator, or while in the pipes, is an impossibility, except in the case of air mixtures which have been discussed earlier in this paper. But the total quantity of these impurities is so very small that the possibility of their ever effecting the ignition of acetylene is an extremely remote one. The instances in which it has ever been claimed that this may have taken place are so few and rest on such slight evidence, that the writer does not believe them at all. Experiments have conclusively shown that the quantity of either one of these necessary to ignite a body of acetylene gas allowed to flow into the air under most favorable conditions, is so great as never to be realized, in the writer's opinion, in the practical use of any carbide likely to be found in the market.

Phosphoretted hydrogen is exceedingly poisonous and most deleterious if inhaled. But when we consider that a small percentage of acetylene in air is sufficient to render the air wholly irrespirable, and that only a few hundredths of a per cent. of the poisonous gas could be present in the acetylene at the most, the possible quantities which a person could take into his lungs in this manner are so small that the writer believes that a case of poisoning by this means could not be brought about, even if the attempt were intentionally made. Certainly no such accident has ever been reported.

Aside from what has been mentioned, no further damage to persons or property would arise from the impurities in acetylene even if quantities of it were to escape constantly into the living apartments of our homes. The ammonia and the sulphuretted hydrogen are soluble in water and will to a varying degree be held back in the water of the generator and of the gas holder. The phosphoretted hydrogen is not so and will all enter the pipes unless removed by a purifier. But supposing all three of them, and these are practically the only ones which need to receive any further attention, should pass through the distributing system and arrive at the burner, what will then become of them? Entering the flame where the other gases are burning, they will, for the most part at least, burn also, and the products of their combustion, like the products of combustion of acetylene itself, will be thrown out into the air. Let us see what these are:

Phosphoretted hydrogen, in burning, produces phosphoric oxide and water vapor.

Sulphuretted hydrogen produces sulphur dioxide and water vapor.

Ammonia produces free nitrogen and water vapor.

Silicon hydride produces silicon dioxide and water vapor.

Of these products the nitrogen, water vapor, and silicon dioxide may be disregarded as entirely harmless. The sulphur dioxide, remaining for a time in the air, may become converted into sulphuric acid and this may be objectionable to health, and especially deleterious to book bindings, draperies, etc. Phosphoric oxide is without a question objectionable in the air and may produce poisonous effects of a serious nature if inhaled in considerable quantity. It is therefore not to be questioned that these impurities in acetylene should be removed from the gas before it arrives at the burner, if they can be present in considerable quantity. Ammonia and hydrogen sulphide are always present in crude coal gas and are removed in the purifying process to which that commodity is always submitted, yet they may be detected in minute quantity in most city gas supplies. The question is one of quantity of impurity and whether we can afford to overlook it because of its very small dimensions.

Purification is discussed in a later topic, but here it may be well to keep before us the fact that the purity of acetylene goes hand in hand with the purity of carbide, and just as prevention is better than cure, so to use only an excellent quality of carbide, thus avoiding impurities, is clearly better than to use a poorer grade, which is sure to result in the formation of these impurities, and then to set about removing them. We are fortunate in the fact that carbide of American manufacture is now so pure that the harmful ingredients never cause any trouble. Enough has been said of impurities arising from carbide, but a word may perhaps be added on the results of overheating.

In the generation of acetylene enormous chemical affinity is displayed between the calcium in the carbide and the oxygen in the water, and as in all such chemical changes, a great quantity of heat is evolved. According to the method of bringing the water and the carbide together, this heat may be localized so as to be harmful, or so distributed as to cause no harm. In the experiment of dropping a fragment of carbide into a tumbler of water, the heat of the reaction is so absorbed by the relatively large body of water that overheating does not occur at any point. But reverse these conditions by allowing water, in limited quantity, drop by drop, to fall on to a lump of carbide and you bring about the overheating of the resultant gas, its conversion to some extent into oily matters, a process known to the chemist as polymerization, and even into a tar. Samples of carbide which have thus been treated in an experimental way have come out of the experiment smeared with a yellow, dirty, tarry mass, which shows evidence at once of overheating in the gas making process, resultant loss of acetylene, and consequently oily matters formed which may cause trouble when the gas laden with them comes along to the burners, causing the latter to clog and refuse to properly operate. These products of overheating are therefore of such a nature as to call for discussion, although they are not impurities in the ordinary sense. They do not pollute the air nor endanger health in any way, but as they cause inconvenience and trouble with the burners, they must not be ignored. Since too great heat, and this alone, causes their formation, the rational remedy is to prevent their production by avoiding high heat at the moment of generation. The remedy is prevention. The means of prevention is *cool generation*.

There is in acetylene another foreign substance, not exactly an impurity, which is often present in small quantities, and may be mechanically carried along from the generator to the distributing system unless arrested. This substance is lime dust or carbide dust. It is easily held back by a simple sifting or filtering arrangement, and the form most frequently used is a filter consisting merely of felt. Every machine should be provided with such a dust arrester or filter, sometimes called a purifier, though its only office is to arrest dust.

COOL GENERATION.

It is essential to emphasize the matter of cool generation as one of the utmost importance in acetylene manufacture. The function of the generator is in principle a simple one. It has to provide for the bringing together of the water and the carbide, wash the gas, filter it as above, purify it if need be, store it to such extent as may be necessary, and deliver it to the pipes for distribution. The more nearly the generator conforms to the simple experiment of dropping a little carbide into a large volume of water, the more efficiently will the heat be distributed and the more perfectly will "cool generation" be accomplished. The other extreme is where a little water is allowed to drip upon an excess of carbide, as in the second experiment; heat is localized instead of distributed, becoming excessive at the point of most rapid generation. Fortunately for the acetylene consumer of the present day, generators which offend most in these particulars have been practically driven out of the market, and the faulty machines of ten years ago are no longer doing business. American generator manufacturers have been so well educated along this line that it is hardly necessary in a work of this sort to longer dwell upon this simple precaution. But since so much trouble came to some of the first generators on this account, it may be well to recall that without care in this direction, and should any manufacturer relax his vigilance so that the opposite of cool generation is secured, in just the same proportion will gas once formed be used up in making the heat products, benzol, styrolene, anthracene, etc.; a lower yield than the correct one will be secured from the carbide; lower candle power will also result, and the tarry products cause carbonization of the burners, resulting in their becoming choked and smoky, three items of great wastefulness, to say nothing of annoyance and inconvenience.

Foremost among the requirements for a perfect, modern generator is, therefore, the one which demands that heat shall be avoided as much as possible. Excessive overheating is detected by the appearance of the lime left after the gas generation. If the residuum is discolored yellow or brown, serious overheating has taken place, for this lime sludge should be essentially white. Tarry deposits in the residue indicate a still greater degree of overheating.

The author learns, by way of one of the German acetylene journals, that an attempt has been made in that country this year to enforce a regulation requiring that the temperature in an acetylene generator shall not exceed two hundred and twelve degrees Fahrenheit. Such a regulation would be entirely too rigid, for while the temperature in the gas containing parts of no good generator ever reaches that point, unless frightfully overworked, nevertheless the momentary temperature of some pieces of carbide, at the instant of reaction, many often exceed two hundred and twelve degrees Fahrenheit, but such strictly local temperatures are quickly quenched by excess of water which must be present. A temperature, even considerably above two hundred and twelve degrees Fahrenheit, may occur within the carbide of any generator; it is inherent in the nature of the reaction, but the remedy is found in abundance of water to serve as a prompt and efficient equalizer.

The German incident is worthy of mention here for two reasons. First, it points out that thinking acetylenists have an eye to the subject of this chapter, *cool generation,* and, secondly, it leads to the reason for a most important regulation in the primer of acetylene which has now become universal and which demands *a gallon of water to every pound of carbide.*

Some earlier generators used only water enough to liberate the gas, and discharged their residuum as a dry ash. The way of these transgressors was hard and we know them no more.

It is well to remember still further, in this connection, that the most prolific cause of overheating is to overwork the generator. If a generator is calculated for ten burners and is called upon to make gas for twenty, the result is sure to be excessive heat production.

TYPES OF GENERATORS.

There are two general methods of bringing water and carbide together for the purpose of generating gas, viz., "carbide to water," and "water to carbide."

Generators are therefore most frequently designated as carbide-feed and water-feed, respectively.

Inasmuch as it is far easier to regulate the flow of liquids, by means of valves and other mechanisms in common use, and universally well-known, than to similarly control the distribution of solid materials, it was very natural that the earlier contrivances were so made as to maintain a properly controlled flow of water, by jets, sprinklers or otherwise, on to quantities of carbide contained in suitable vessels. Later, because of the very obvious fact that the most rational plan is to drop carbide, in suitable quantities, into a large excess of water, mechanisms in some variety were developed to feed the solid into the liquid. As the years have advanced all the eccentric and miscellaneous modifications have practically vanished and only the fittest of both types have survived.

1. Water-feed Machines. The modern water-to-carbide generator usually consists of two parts, the generator proper and the gas holder. Carbide is charged into what is known as the divided pan, a tray divided into segments or compartments containing about two pounds in each compartment and so arranged that the water first attacks the carbide in compartment 1, exhausts and thoroughly floods it, then overflows into compartment 2, where it finds a fresh supply of carbide. This being exhausted and thoroughly flooded,

the water may then pass to 3, and so on till all the cells are flooded in rotation, when the advancing water passes from the last compartment into cell 1 of another precisely similar pan below, and the process is repeated, the pans being always surrounded by cold water.

This type of generator is especially adapted to localities where running water is not available and where it is not possible or convenient to connect with ample drainage pipes for removal of sludge and for flushing.

These generators also have the advantage of using large sizes of carbide; the refuse is carried away in the pans and water brought to the machine in buckets.

2. Carbide-feed Machines. The modern carbide-to-water generators also consist of two parts, the generator proper and the gas-holder or bell. The generator in all these machines contains a hopper of some sort which receives the carbide, fitted with a mechanism to control it and admit of its dropping into the water below at the right times and in small, regulated quantities. Such mechanisms are of two kinds; one depends upon releasing the carbide by some sort of a valve whereby it may drop, directly or indirectly, into the water below (gravity feed); the other depends on pushing the carbide off a plate or some similar device, usually by means of clock work (forced feed). Whether an apparatus is based on one or the other of these principles, the carbide being sustained in a receptacle above the water, it is imperative that its movement must be so safeguarded that it is impossible for the machine to drop its charge uncontrolled, into the water, either by accident or by intent. The feeding device, of whichever type it may be, is obviously the most critical and vital part of a carbide feed mechanism. On it depends uninterrupted and perfect lighting service as well as safety and economy in operation. It should be positive, strong and simple; must not get out of order; must respond promptly to every demand made upon it and never feed either too much or too little carbide into the generating chamber. While it must release carbide whenever carbide is called for, it must with equal reliability prevent the feeding of carbide at all times when it is not called for. Since moisture acts on carbide in such a way as to render it in time more or less sticky and less easily responsive to mechanical movements, especially after a charged generator has been standing for some time inactive, it follows that the feeding device must also be, so far as possible, self-cleaning.

Carbide feed generators usually require a special size of carbide "Nut," 1¼x⅜ or "Quarter," ¼x$^{1}/_{12}$ being in commonest use.

In either water-to-carbide or carbide-to-water generators, the gas-holder, or bell, performs an important function, that of regulating the feed; it must do this with positiveness and precision. The fall of the bell, indicating as it does, approaching exhaustion of gas supply, releases the mechanism, whatever it may be, which starts the feeding; the accumulated gas raises the bell, and the rising bell must arrest the feed. In a nicely adjusted generator, the rise and fall of the bell is very slight, a slow feeding and consequently slow generation of gas assuring cool generation and steady pressure.

Of whatever type, every generator must have sufficient holder capacity to provide for all generation in excess of use at any time, including what is called "after generation," which is the amount of gas generated after the lights are turned off, but before gas evolution can entirely cease. This is usually accomplished by the well known bell gas-holder as just described; there are other devices to accomplish the same purpose, but the bell is so common as to be almost universally present in the machines available in this country. Water-feed generators are usually equipped with greater bell capacity than carbide-feed, because in this type of apparatus more surface of carbide is exposed to moist air, hence a greater amount of after generation occurs.

Acetylene generators are thus seen to be automatic, that is, aside from periodical care in cleaning and recharging, they are designed to need no attention. They provide for the automatic evolution of gas as fast as gas is drawn at the burners, and to cut off the liberation of gas, also automatically, so soon as practicable after consumption ceases. Large installations, such as for village lighting, extensive factories, etc., are properly so constructed as to call for frequent attention during the time when the gas generation takes place, and large holders are used to maintain a supply; but small plants, such as are employed for household use, ought to demand no attention other than that of occasional charging and removing of residue, in other words, must automatically supply water to carbide, or carbide to water, as the case may be, during the time in which gas is being used. It would, of course, be possible to generate at certain intervals when the time of an attendant could be given to this duty, supplying carbide by hand to the water, and storing enough gas to last for a day or two. But this method demands unnecessarily large gasometers, which are bulky, expensive and difficult to place, hence this method is nowhere in vogue for small installations, the automatic machines being everywhere preferred.

Of whatever type it may be, a good generator, such as can be recommended for household use, must be possessed of certain qualities which will be conceded by all:

(1) It must allow no possibility of the existence of an explosive mixture in any of its parts at any time. It is not enough to argue

that a mixture, even if it exists, cannot be exploded unless kindled. It is necessary to demand that a dangerous mixture can at no time be formed even if the machine is tampered with by an ignorant person. The perfect machine must be so constructed that it shall be impossible at any time under any circumstances to blow it up. It must be "fool-proof."

(2) It must ensure cool generation. Since this is a relative term, all machines being heated somewhat during the generation of gas, this amounts to saying that a machine must heat but little. A pound of carbide decomposed by water develops the same amount of heat under all circumstances, but that heat can be allowed to increase locally to a high point, or it can be equalized by water so that no part of the material becomes heated enough to do damage.

(3) It must be well constructed. A good generator does not need, perhaps, to be "built like a watch," but it should be solid, substantial, of good material. No light weight, half-price metal, likely to rust through and cause leakage and resultant gas mixtures, should be tolerated. It should be built of the best material adapted to the purpose. It should be built for service, to last and not simply to sell; anything short of this is to be avoided as unsafe and unreliable.

(4) It must be simple. The more complicated the machine the sooner it will get out of order. Understand your generator. Know what is inside of it and beware of an apparatus, however attractive its exterior, whose interior is filled with pipes and tubes, valves and diaphragms, whose functions you do not perfectly understand. If a complicated mechanism is employed to perform what seems to you a simple duty, rely upon your own common sense and look further till you find a perfectly simple but strong mechanism to perform the work of automatically making gas. There are plenty of them and you can afford to meet the price of the machine which is least likely to call for repair next season.

(5) It should create no considerable pressure in any of its parts. More than a pound of pressure at any point may be a source of danger; more than a few ounces is wholly unnecessary, and not to be tolerated.

(6) It should be capable of being cleaned and recharged and receiving all other necessary attention without loss of gas, both for economy's sake, and more particularly to avoid filling the house with a disagreeable odor. There is no need of any perceptible odor about the machine, or in the house, and the better machines successfully guard against this nuisance.

(7) It should require little attention. All machines have to be emptied and recharged periodically; but the more this process is simplified and the more quickly this can be accomplished, the better.

(8) It should be provided with a suitable indicator to designate how low the charge is in order that the refilling may be done in good season. A generator which can by any reasonable possibility leave the household, without warning, plunged in darkness, is not to be tolerated.

(9) It should completely use up the carbide, generating the maximum amount of gas.

As to the selection of type of generator, there are good generators of both types, and under local conditions it should be always an easy matter to select the most suitable one. The famous Moissan, French savant and chemist, wrote as long ago as December, 1896, as follows:

"The ideal apparatus, which I think does not yet exist, consists in a gas holder containing an excess of water into which a fragment of carbide of known weight falls automatically at the desired moment. The weight of this carbide should be such that it will fill the gas holder with gas without producing an excess. Moreover, this carbide should not fall till the moment when the gas holder is nearly empty."

And Professor Lewes, English chemist and illuminating gas authority, wrote early in 1898:

"The generators of the carbide-to-water type are undoubtedly the best, as, with the water kept in excess, it is impossible for the temperature to rise above the boiling point of water, and under all conditions this class of generators yields the purest gas, as the acetylene, having to bubble through the lime water, formed in the generator, is washed free from most of its impurities." Also, still referring to generators of the same class, "With a properly arranged tank the temperature never exceeds the air temperature by more than a few degrees. Under these conditions, the absence of polymerization and the washing of the nascent and finely divided bubbles of gas by the lime water, in the generator, yields acetylene of a degree of purity unapproached in any other form of apparatus."

In following the development of generators in this country from the earliest types to the end of 1908, the writer thinks that the ideals of Moissan have been more than attained in the modern carbide feed generators of the present day, and that the best water-feed machines which are now to be had cover all the points of rational generation so perfectly that, in places where this type is more perfectly adjusted to the surroundings and conveniences, there need be no hesitation in using them.

SUPPLY OF CARBIDE.

The Union Carbide Company, of Chicago and New York, operating under patents, manufactures and distributes the supply of calcium carbide for the entire United States and its dependencies. Plants for the production of calcium carbide have been established for several years at Niagara Falls, N. Y., and Sault Ste. Marie, Mich.

The extraordinary advance in the consumption of carbide has made necessary, repeated and, in the aggregate, enormous additions to the productive capacity of the company. For several years, however, the supply has been, and for the future will be, adequate to meet all domestic demands.

The Union Carbide Company, by a system of established warehouses in some fifty cities, selected with direct reference to the convenience of the consumer, has enabled the user of calcium carbide to secure his supply almost immediately, no matter where he may be located. This system of local warehouses is being as rapidly extended as necessity requires and has proved of great satisfaction and advantage to the small purchaser.

For several years after this company commenced operations its plan was to sell calcium carbide in car-lots to various dealers throughout the country, many of whom were agents for, or manufacturers of, acetylene generators. Under this plan, the price of calcium carbide to the consumer was materially higher than at present, in order to allow a margin of profit to the dealer, and the cost of carbide varied all over the country, depending to a great extent upon freight rates.

It was soon seen that the best interests of the consumer would be served by the establishment of a uniform price, which could only be accomplished by the prepayment of freight to warehouses located in the near vicinity of the consumer, and thus cutting out completely all profits between the manufacturer and the purchaser. This change of method of distribution has resulted, in the last few years, in reductions in the price of calcium carbide of as much as twenty-five dollars per ton in some localities. The United States possesses, therefore, a distinct advantage in this arrangement of the business, and as a strictly uniform carbide of the highest grade is produced, the user of acetylene is thereby enabled to assure himself as to the quality of every pound of calcium carbide which he purchases.

There are in operation elsewhere throughout the world numerous small plants manufacturing calcium carbide, but in foreign countries the warehouse system is rendered impossible, transportation charges from the manufacturer to the consumer are at local and often extravagant rates, and confusion as to quality creates a difficulty unknown in the United States.

The price of carbide in the United States is established on the basis of all freight prepaid to the local warehouse, thus avoiding the long haul and the very much higher expense attached to the method of distribution in foreign countries. Thus, while the price at the works in Europe approximates the price at the warehouse in the United States, the freight to the point of delivery must be added when comparisons are made. The prevailing price of calcium carbide in foreign countries is somewhat higher, and the quality of the carbide considerably lower than in the United States.

The carbide industry has decidedly outgrown its infancy, and is now a highly organized, scientific business conducted on the broadest of sound principles.

TRANSPORTATION AND STORAGE OF CARBIDE.

Carbide, to be properly preserved for storage and shipment, must be packed as soon as made in some sort of hermetically sealed containers to preserve it from the deteriorating action of moisture. When dry, it is a perfectly harmless substance, being solid and incombustible. Its only dangerous attributes can be developed solely by contact with water. Well packed, it is as safe to store or transport as other ordinary commodities. If the building in which it is stored should burn, unbroken packages will go through the fire unharmed, so also would loose lumps if packages break, except for the circumstance that considerate firemen usually play plenty of water upon them, compelling them to generate gas, which in turn burns and contributes to the conflagration. But in itself there is nothing about carbide to render it to any extent a hazardous article. Very early in its history certain false impressions were disseminated about carbide, and have by their persistence much handicapped this deserving though youthful aspirant for honors in the field of industry.

As the London Chemical Trade Journal put it, "All trouble has, we think, undoubtedly arisen through a certain amount of misunderstanding and confusion of ideas. It would seem that the fire insurance people have not been quite clear as to the difference in the nature of calcium carbide and acetylene. Knowing that the one generally had something to do with the other, they have evidently thought they were pretty much of a muchness, and made no distinction." But all these things have passed away and the notion that calcium carbide is a dangerous material for transportation is a relic of the past.

Calcium carbide is packed for shipment and delivery in various sized sheet steel drums, containing from fifty to one thousand pounds each; the price includes the cost of the drums, which, being used but once, are not likely to become so battered as to prevent their remaining air-tight.

Following is a reprint of a circular issued by Department of Commerce and Labor:

TRANSPORTATION OF CALCIUM CARBIDE ON PASSENGER STEAMERS.

Department of Commerce and Labor. Steamboat Inspection Service.

WASHINGTON, February 28, 1907.

Bureau Circular No. 2.

To Supervising and Local Inspectors of Steamboats, Owners of Steamboats, and Others Concerned:

Treasury Decision No. 19507, dated June 18, 1898, in so far as the same prohibits the transportation of carbide of calcium or calcium carbide on passenger steamers, is hereby rescinded, and such transportation, from and after the date hereof, is hereby permitted, provided the following requirements are observed:

Calcium carbide or carbide of calcium may be transported on passenger vessels when the same is contained in steel drums or steel receptacles, the seams of which are lapped and properly riveted or fastened in such manner as will best insure the maximum strength of the joints, and when the said drums or receptacles are fitted with double covers, so that such drums or receptacles shall be watertight and airtight during such transportation. For packages of one hundred and ten pounds or less, such drums or receptacles shall be made of open-hearth steel of not less than No. 26 gauge. For packages of more than one hundred and ten pounds, such drums or receptacles shall be made of open-hearth steel of not less than No. 24 gauge: Provided, that said calcium carbide or carbide of calcium may also be transported on passenger vessels in cans containing not more than ten pounds of said calcium carbide or carbide of calcium each, such cans to be constructed and packed as follows: Said cans shall be made of ninety-pound basis tin or terneplate, shall have double and lapped seams, shall have the openings in the tops fitted with tight compression caps, shall be constructed so as to render them watertight and airtight, and shall be packed in substantial wooden cases with not more than sixty pounds total weight of calcium carbide or carbide of calcium in each of such cases.

Calcium carbide or carbide of calcium must not under any conditions be transported on such vessels in other than such original unopened packages, which must be plainly marked "Calcium carbide."

GEO. UHLER, Supervising Inspector-General.

Approved: OSCAR S. STRAUS, Secretary.

INSURANCE REGULATIONS.

It would lead beyond the intended limits of this bulletin to enter upon a discussion of insurance regulations which have existed in this or other countries, but which are not now in force. Those in operation in this country at the present time, annulling, as of course they do, all previous rules, are alone of interest to us. All insurance regulations affecting carbide and acetylene should be based on the most accurate scientific knowledge of the gas and the method of its production. The National Board of Fire Underwriters have adopted a code of rules and regulations for the construction, installation and use of acetylene machines, and for the storage of calcium carbide, as recommended by its Committee of Consulting Engineers, who have elaborated the matter with scrupulous care. These regulations have been amended to June, 1908. They first of all *absolutely prohibit the use of liquefied acetylene and of gas generated therefrom.*. Part 1 of the code contains rules for the installation of generators, part 2 embraces regulations for the storage of carbide, and part 3 for the construction of generators. Generators are grouped in two classes, A and B. Class A embraces all stationary automatic apparatus, class B stationary non-automatic generators.

Such parts of the rules as are of particular interest for the purposes of this bulletin will be here given in full:

PART 1.

RULES FOR THE INSTALLATION AND USE OF ACETYLENE GAS GENERATORS.

The use of liquid acetylene or gas generated therefrom is absolutely prohibited.

Failure to observe these rules is as liable to endanger life as property.

To secure the largest measure of safety to life and property, the following rules for the installation of acetylene gas machines must be observed.

CLASS A.

Stationary Automatic Apparatus.

1. Foundations. a. Must, where practicable, be of brick, stone, concrete or iron. If necessarily of wood they shall be extra heavy, located in a dry place and open to the circulation of air.

The ordinary board platform is not satisfactory. Wooden foundations shall be of heavy planking, joists or timbers, arranged so that the air will circulate around them and so as to form a firm base.

b. Must be so arranged that the machine will be level and unequal strain will not be placed on the generator or connections.

2. Location. a. Generators, especially in closely built up districts, should preferably be placed outside of insured buildings, in generator houses constructed and located in compliance with Rule 9.

b. Generators must be so placed that the operating mechanism will have room for free and full play and can be adjusted without artificial light. They must not be subject to interference by children or careless persons, and if for this purpose further enclosure is necessary, it must be furnished by means of slatted partitions, permitting the free circulation of air.

c. Generators which from their construction are rendered inoperative during the process of recharging must be so located that they can be recharged without the aid of artificial light.

d. Generators must be placed where water will not freeze.

3. Escapes or Relief Pipes. Each generator must be provided with an escape or relief pipe of ample size; no such pipe to be less than three-quarters inch internal diameter. This pipe shall be substantially installed, without traps, and so that any condensation will drain back to the generator. It must be carried to a suitable point outside the building, and terminate in an approved hood located at least twelve feet above ground and remote from windows.

The hood must be constructed in such a manner that it cannot be obstructed by rain, snow, ice, insects or birds.

4. Capacity. a. Must be sufficient to furnish gas continuously for the maximum lighting period to all lights installed. A lighting period of at least five hours shall be provided for in every case.

b. Generators for conditions of service requiring lighting periods of more than five hours must be of sufficient capacity to avoid recharging at night. The following ratings will usually be found advisable.

(1) For dwellings, and where machines are always used intermittently, the generator must have a rated capacity equal to the total number of burners installed.

(2) For stores, opera houses, theatres, day run factories, and similar service, the generator must have a rated capacity of from thirty to fifty per cent. in excess of the total number of burners installed.

(3) For saloons and all night or continued service, the generator must have a rated capacity of from one hundred to two hundred per cent. in excess of the total number of burners installed.

c. A small generator must never be installed to supply a large number of lights, even though it seems probable that only a few lights will be used at a time. *An overworked generator adds to the cost of producing acetylene gas.*

5. Carbide Charges. Must be sufficient to furnish gas continuously for the maximum lighting period to all burners installed. In determining charges lump carbide must be estimated as capable of producing four and one-half cubic feet of gas to the pound, commercial one-fourth inch carbide four cubic feet of gas to the pound, and burners must be considered as requiring at least twenty-five per cent. more than their rated consumption of gas.

6. Burners. Burners consuming one-half of a cubic foot of gas per hour are considered standard in rating generators. Those having a greater or less capacity will decrease or increase the number of burners allowable in proportion.

Burners usually consume from twenty-five to one hundred per cent. more than their rated consumption of gas depending largely on the working pressure. The so-called ½-foot burner when operated at pressures of from 20 to 25 tenths inches water column (2 to 2½ inches) is usually used with best economy.

7. Piping. a. Connections from generators to service pipes must be made with right and left thread nipples or long thread nipples with lock nuts. All forms of unions are prohibited.

b. Piping must, as far as possible, be arranged so that any moisture will drain back to the generator. If low points occur of necessity in any piping, they must be drained through tees into drip cups permanently closed with screw caps or plugs. No pet-cocks shall be used.

c. A valve and by-pass connection must be provided from the service-pipe to the blow-off for removing the gas from the holder in case it should be necessary to do so.

d. The schedule of pipe sizes for piping from generators to burners should conform to that commonly used for ordinary gas, but in no case must the feeders be smaller than three-eights inch.

The following schedule is advocated:

⅜	inch pipe,	26 feet,	three burners.
½	inch pipe,	30 feet,	six burners.
¾	inch pipe,	50 feet,	twenty burners.
1	inch pipe,	70 feet,	thirty-five burners.
1¼	inch pipe,	100 feet,	sixty burners.
1½	inch pipe,	150 feet,	one hundred burners.
2	inch pipe,	200 feet,	two hundred burners.
2½	inch pipe,	300 feet,	three hundred burners.
3	inch pipe,	450 feet,	four hundred and fifty burners.
3½	inch pipe,	500 feet,	six hundred burners.
4	inch pipe,	600 feet,	seven hundred and fifty burners.

e. Machines of the carbide-feed type must not be fitted with continuous drain connections leading to sewers, but must discharge into suitable open receptacles which may have such connections.

f. Piping must be thoroughly tested before and after the burners have been installed. It must not show loss in excess of two inches within twelve hours when subjected to a pressure equal to that of fifteen inches of mercury.

g. Piping and connections must be installed by persons experienced in the installation of acetylene apparatus.

8. Care and Attendance. In the care of generators designed for a lighting period of more than five hours, always clean and re-charge the generating chambers at regular stated intervals regardless of the number of burners actually used.

Where generators are not used throughout the entire year, always remove all water and gas and clean thoroughly at the end of the season during which they are in service.

It is usually necessary to take the bell portion out and invert it so as to allow all gas to escape. This should never be done in the presence of artificial light or fire of any kind.

Always observe a regular time, during daylight hours only, for attending to and charging the apparatus.

In charging the generating chambers of water-feed machines, clean all residuum carefully from the containers and remove it at once from the building. Separate from the mass any unslacked carbide remaining and return it to the retainers, adding new carbide as required. Be careful never to fill the containers over the specified mark, as it is important to allow for the swelling of the carbide when it comes in contact with water. The proper action and economy of the machine are dependent on the arrangement and amount of carbide placed in the generator. Carefully guard against the escape of gas.

Whenever re-charging with carbide always replenish the water supply.

Never deposit residuum or exhausted material from water-feed machines in sewer pipes or near inflammable material.

Always keep water tanks and water seals filled with clean water.

Never test the generator or piping for leaks with a flame, and never apply flame to an outlet from which the burner has been removed.

Never use a lighted match, lamp, candle, lantern or any open light near the machine.

Failure to observe the above cautions is as liable to endanger life as property.

OUTSIDE GENERATOR HOUSES.

9. a. Outside generator houses should not be located within five (5) feet of any opening into, nor shall they open toward any adjacent building, and must be kept under lock and key.

b. The dimensions must be no greater than the apparatus requires to allow convenient room for recharging and inspection of parts. The floor must be at least twelve (12) inches above grade and the entire structure thoroughly weather-proof.

c. Generator houses must be thoroughly ventilated, and any artificial heating necessary to prevent freezing shall be done by steam or hot water systems.

d. Generator houses must not be used for the storage of calcium carbide except in accordance with the rules relating to that subject.

CLASS B.

Stationary Non-Automatic Apparatus.

10. Foundations. a. Must be of brick, stone or concrete.

b. Must be so arranged that the machine will be level and so that strain will not be brought upon the connections.

11. Gas Houses. a. Must be constructed entirely of non-combustible material and must not be lighted by any system of illumination involving open flames.

b. Must be heated, where artificial heating is necessary to prevent freezing, by steam or hot water systems, the heater to be located in a separate building, and no open flames to be permitted within generator enclosures.

c. Must be kept closed and locked excepting during daylight hours.

d. Must be provided with a permanent and effective system of ventilation which will be operative at all times, regardless of the periods of operation of the plant.

12. Escape Pipes. Each generator must be provided with a vent pipe of ample size, substantially installed, without traps. It must be carried to a suitable point outside the building and terminate in an approved hood located at least twelve feet above ground and remote from windows.

The hood must be constructed in such a manner that it cannot be obstructed by rain, snow, ice, insects or birds.

13. Care and Maintenance. All charging and cleaning of apparatus, generation of gas and execution of repairs must be done during daylight hours only, and generators must not be manipulated or in any way tampered with in the presence of artificial light.

This will require gas-holders of a capacity sufficient to supply all lights installed for the maximum lighting period, without the necessity of generation of gas at night or by artificial light.

In the operation of generators of the carbide-feed type it is important that only a limited amount of carbide be fed into a given body of water. An allowance of at least one gallon of generating water per pound of carbide must be made in every case, and when this limit has been reached the generator should be drained and flushed, and clean water introduced. These precautions are necessary to avoid overheating during generation and accumulation of hard deposits of residuum in the generating chamber.

Part 2.

RULES FOR THE STORAGE OF CALCIUM CARBIDE.

14. Storage of Calcium Carbide. a. Calcium Carbide in quantities not to exceed six hundred (600) pounds may be stored, when contained in approved metal packages not to exceed one hundred (100) pounds each, inside insured property, provided that the place of storage be dry, waterproof and well ventilated, and also provided that all but one of the packages in any one building shall be sealed and the seals shall not be broken so long as there is carbide in excess of one (1) pound in any other unsealed package in the building.

b. Calcium Carbide in quantities in excess of six hundred (600) pounds must be stored above ground in detached buildings, used exclusively for the storage of Calcium Carbide, in approved metal packages, and such buildings shall be constructed to be dry, waterproof and well ventilated.

c. Packages to be approved must be made of metal of sufficient strength to insure handling the package without rupture, and be provided with a screwed top or its equivalent.

They must be constructed so as to be water and air-tight without the use of solder, and conspicuously marked "CALCIUM CARBIDE—DANGEROUS IF NOT KEPT DRY."

Part 3.

RULES FOR THE CONSTRUCTION OF GENERATORS.

The following rules are intended to provide only against the more hazardous defects usually noted in apparatus of this kind. The rules do not cover all details of construction nor the proper proportioning of parts, and devices which comply with these requirements alone are not necessarily suitable for listing as permissible for use. These points are often only developed in the examination required before permission is given for installation.

CLASS A.

STATIONARY APPARATUS FOR ISOLATED INSTALLATIONS.

15. General Rules—Generators. a. Must be made of iron or steel, and in a manner and of material to insure stability and durability.

b. Must be automatically regulated and uniform in their action, producing gas only as immediate consumption demands, and so designed that gas is generated without producing sufficient heat to cause yellow discoloration of residuum (which will occur at about five hundred degrees Fahrenheit) or abnormal pressure at any stage of the process when using carbide of any degree of fineness.

The presence of excessive heat tends to change the chemical character of the gas and may even cause its ignition, while in machines of the carbide-feed type, finely divided carbide will produce excessive pressure unless provision is made to guard against it.

c. Must be so arranged that during recharging back flow of gas from the gas-holder will be automatically prevented, or so arranged that it will be impossible to charge the apparatus without first closing the supply pipe to the gas-holder, and to the other generating chambers if several are used.

This is intended to prevent the dangerous escape of gas.

d. The water or carbide supply to the generating chamber must be so arranged that gas will be generated long enough in advance of the exhaustion of the supply already in the gas-holder to allow the using of all lights without exhausting such supply.

This provides for the continuous working of the apparatus under all conditions of water feed and carbide charge, and it obviates the extinction of lights through intermittent action of the machine.

e. No valves or pet-cocks opening into the room from the gas-holding part or parts, the draining of which will allow an escape of gas, are permitted, and condensation from all parts of the apparatus must be automatically removed without the use of valves or mechanical working parts.

Such valves and pet-cocks are not essential; their presence increases the possibility of leakage. The automatic removal of condensation from the apparatus is essential to the safe working of the machine.

U-traps opening into the room from the gas-holding parts must not be used for removal of condensation. All sealed drip connections must be so arranged as to discharge gas to the blow-off when blown out, and the seals must be self-restoring upon relief of abnormal pressure.

The apparatus must be capable of withstanding fire from outside causes.

Sheet metal joints must be double seamed or riveted and thoroughly sweated with solder. Pipes must be attached to sheet metal with lock-nuts or riveted flanges.

This prohibits the use of wood or of joints relying entirely upon solder.

g. Gage glasses, the breakage of which would allow the escape of gas must not be used.

h. The use of mercury seals is prohibited.

Mercury has been found unreliable as a seal in acetylene apparatus.

i. Combustible oils must not be used in connection with the apparatus.

j. The construction must be such that liquid seals shall not become thickened by the deposit of lime or other foreign matter.

k. The apparatus must be constructed so that accidental siphoning of water will be impossible.

l. Flexible tubing, swing joints, unions, springs, mechanical check valves, chains, pulleys, stuffing boxes and lead or fusible piping must not be used on acetylene apparatus, except where failure of such parts will not vitally affect the working or safety of the machine.

Floats must not be used excepting in cases where failure will result only in rendering the machine inoperative.

m. Every machine must be plainly marked with the maximum number of lights it is designed to supply, the amount of carbide necessary for a single charge, the manufacturer's name and the name of the machine.

16. *Generating Chambers.* a. Must be constructed of galvanized iron or steel not less than No. 24 U. S. Standard gage in thickness for capacities up to and including 20 gallons, not less than No. 22 U. S. Standard gage for capacities between 20 and 75 gallons, and not less than No. 20 U. S. Standard gage for capacities in excess of 75 gallons.

b. Must each be connected with the gas-holder in such a manner that they will, at all times, give open connection either to the gas-holder or to the blow-off pipe to the outer air.

This prevents dangerous pressure within or the escape of gas from the generating chamber.

c. Must be so constructed that not more than 5 pounds of carbide can be acted upon at once, in machines which apply water in small quantities to the carbide.

This tends to reduce the danger of over-heating and excessive after-generation by providing for division of the carbide charges in machines of this type.

d. Must be provided with covers having secure fastenings to hold them properly in place, and those relying on a water seal must be submerged in at least twelve inches of water. Water seal chambers for covers depending on a water seal must be one and one-half inches wide and fifteen inches deep, excepting those depending upon the filling of the seal chambers for the generation of gas, where nine inches will be sufficient.

e. Must be so designed that the residuum will not clog or affect the working of the machine and can conveniently be handled and removed.

f. Must be provided with suitable vent connections to the blow-off pipe so that residuum may be removed and the generating water replaced without causing siphoning or introducing air to the gas-holder upon recharging.

This applies to machines of the carbide-feed type.

g. Feed mechanism for machines of the carbide feed type must be so designed that the direct fall of carbide from the carbide holder into the water of the generator is prevented at all positions of the feed mechanisms; or, when actuated by the rise and fall of a gas bell, must be so arranged that the feed valve will not remain open after the landing of the bell, and so that the feed valve remains inoperative as long as the filling opening on the carbide hopper remains open. Feed mechanisms must always be far enough above the water level to prevent clogging from the accumulation of damp lime. For this purpose the distance should be not less than 10 inches.

17. *Carbide Chambers.* a. Must be constructed of galvanized iron or steel not less than No. 24 U. S. Standard gage in thickness for capacities up to and including fifty (50) pounds and not less

than No. 22 U. S. Standard gage for capacities in excess of fifty (50) pounds.

b. Must have sufficient carbide capacity to supply the full number of burners continuously and automatically during the maximum lighting period.

This rule removes the necessity of recharging or attending to the machine at improper hours. Burners almost invariably require more than their rated consumption of gas and carbide is not of staple purity, and there should therefore be an assurance of sufficient quantity to last as long as light is needed. Another important consideration is that in some establishments burners are called upon for a much longer period of lighting than in others, requiring a generator of greater gas producing capacity. Machines having several generating chambers must automatically begin generation in each upon exhaustion of the preceding chamber.

c. Must be arranged so that the carbide holders or charges may be easily and entirely removed in case of necessity.

18. *Gas Holders.* a. Must be constructed of galvanized iron or steel not less than No. 24 U. S. Standard gage in thickness for capacities up to and including 20 gallons, not less than No. 22 U. S. Standard gage for capacities between 20 and 75 gallons, and not less than No. 20 U. S. Standard gage for capacities in excess of 75 gallons.

Gas bells, if used, may be two gages lighter than holders.

Condensation chambers, if placed under holders, to be of same gage as holders.

b. Must be of sufficient capacity to contain all gas generated after all lights have been extinguished.

If the holder is too small and blows off frequently after the lights are extinguished there is a waste of gas. This may suggest improper working of the apparatus and encourage tampering.

c. Must, when constructed on the gasometer principle, be so arranged that when the gas bell is filled to its maximum with gas at normal pressure, its lip or lower edge will extend at least 9 inches below the inner water level.

d. Must, when constructed on the gasometer principle, have the dimensions of the tank portion so related to those of the bell that a pressure of at least eleven (11) inches will be necessary before gas can be forced from the holder.

e. The bell portion of a gas-holder constructed on the gasometer principle must be provided with a substantial guide to its upward movement, preferably in the center of the holder, carrying a stop acting to check the bell one inch above the normal blow-off point.

This tends to insure the proper action of the bell and decreases the liability of escaping gas.

f. A space of at least three-quarters of an inch must be allowed between the sides of the tank and the bell.

g. All water seals must be so arranged that the water level may be readily seen and maintained.

19. *Water Supply.* a. The supply of water to the generator for generating purposes must not be taken from the water seal of any gas-holder constructed on the gasometer principle, unless the feed mechanism is so arranged that the water seals provided for in Rules 20, b, c and d, may be retained under all conditions.

This provides for the proper level of water in the gas-holder.

b. In cases where machines of the carbide-feed type are supplied with water from city water mains or house pipes, the pipe connection must discharge into the regularly provided filling trap on the generator and not through a separate continuous connection leading into the generating chamber.

This is to prevent the expulsion of explosive mixtures through the filling trap in refilling.

20. *Reliefs or Safety Blow-offs.* a. Must in all cases be provided, and must afford free vent to the outer air for any overproduction of gas, and also afford relief in case of abnormal pressure in the machine.

Both the above mentioned vents may be connected with the same escape pipe.

b. Must be at least ¾-inch internal diameter and be provided with suitable means for connecting to the pipe leading outside of the building.

c. Must be constructed without valves or other mechanical working parts.

d. Apparatus requiring pressure regulators must be provided with an additional approved safety blow-off attachment located between the pressure regulator and the service pipes and discharging to the outer air.

This is intended to prevent the possibility of under pressure in the service pipes due to failure of the pressure regulator.

21. *Pressures.* a. The working pressure at the generator must not vary more than ten-tenths (one) inch water column under all conditions of carbide charge and feed, and between the limits of no load and 50 per cent. overload.

b. Apparatus not requiring pressure regulators must be so arranged that the gas pressure cannot exceed sixty-tenths (six) inches water column.

This requires the use of the pressure relief provided for in Rule No. 20a.

c. Apparatus requiring pressure regulators must be so arranged that the gas pressure cannot exceed three pounds to the square inch.

The pressure limit of three pounds is taken since that is the pressure corresponding to a water column about six feet high, which is about the limit in point of convenience for water sealed reliefs.

22. *Air Mixtures.* Generators must be so arranged as to contain the minimum amount of air when first started or recharged, and no device or attachment facilitating or permitting mixture of air with the gas prior to consumption, except at the burners, shall be allowed.

Owing to the explosive properties of acetylene mixed with air, machines must be so designed that such mixtures are impossible.

23. *Purifiers.* a. Must be constructed of galvanized iron or steel not less than No. 24. U. S. Standard gage in thickness.

b. Where installed, purifiers must conform to the general rules for the construction of other acetylene apparatus and allow the free passage of gas.

c. Purifiers must contain no carbide for drying purposes.

d. Purifiers must be located inside of gas-holders, or, where necessarily outside, must have no hand-holes which can be opened without first shutting off the gas supply.

24. *Pressure Regulators.* a. Must conform to the rules for the construction of other acetylene apparatus so far as they apply and must not be subject to sticking or clogging.

b. Must be capable of maintaining a uniform pressure, not varying more than four-tenths inch water column, at any load within their rating.

c. Must be installed between valves in such a manner as to facilitate inspection and repairs.

CLASS B.

Stationary Apparatus for Central Station Service.

Generators of over 300 lights capacity for central station service are not required to be automatic in operation. Generators of less than 300 lights capacity must be automatic in operation and must comply in every respect with the requirements of Part 3, Class A.

25. *General Rules—Generators.* a. Must be substantially constructed of iron or steel and be protected against depreciation by an effective and durable preventive of corrosion.

Galvanizing is strongly recommended as a protection against oxidation, and it may to advantage be reinforced by a thorough coating of asphaltum or similar material.

b. Must contain no copper or alloy of copper in contact with acetylene, excepting in valves.

c. Must be so arranged that generation will take place without overheating; temperature in excess of 500° F. to be considered excessive.

d. Must be provided with means for automatic removal of condensation from gas passage.

e. Must be provided with suitable protection against freezing of any water contained in the apparatus.

No salt or other corrosive chemical is permissible as a protection against freezing.

f. Must in general comply with the requirements governing the construction of apparatus for isolated installations so far as they are applicable.

g. Must be so arranged as to insure correct procedure in recharging and cleaning.

h. Generators of the carbide-feed type must be provided with some form of approved measuring device to enable the attendant to determine when the maximum allowable quantity of carbide has been fed into the generating chamber.

In the operation of generators of this type an allowance of at least one gallon of clean generating water per pound of carbide should be made, and the generator should be cleaned after slacking of every full charge. Where lump carbide is used the lumps may become embedded in the residuum, if the latter is allowed to accumulate at the bottom of the generating chamber, causing overheating from slow and restricted generation, and rendering the mass more liable to form a hard deposit and bring severe stress upon the walls of the generator by slow expansion.

26. *Generating Chambers.* a. Must each be connected with the gas-holder in such a manner that they will, at all times, give open connection either to the gas-holder or to the blow-off pipe into the outer air.

b. Must be so arranged as to guard against appreciable escape of gas to the room at any time during the introduction of the charges.

c. Must be so designed that the residuum will not clog or affect the operation of the machine, and can conveniently be handled and removed.

d. Must be so arranged that during the process of cleaning and recharging, backflow of gas from the gas-holder or other generating chambers will be automatically prevented.

27. *Gas-Holders.* a. Must be of sufficient capacity to contain at least 4 cubic feet of gas per ½-foot burner of the rating.

This is to provide for the requisite lighting period without the necessity of making gas at night, allowance being made for the enlargement of burners caused by the use of cleaners.

b. Must be provided with suitable guides to direct the movement of the bell throughout its entire travel.

28. *Pressure Reliefs.* Must in all cases be provided, and must be so arranged as to prevent pressure in excess of 100 tenths (10) inches water column in the mains.

29. *Pressures.* Gas-holder must be adjusted to maintain a pressure of approximately 25 tenths (2.5) inches water column in the mains.

CLASS D.

Portable Table Lamps.

30. *General Rules—Lamps.* a. Must be substantially made of metal, and the construction must embody no copper, either pure or alloyed, in contact with acetylene.

b. Must in all parts subject to corrosion be thoroughly protected by an effective and durable preventive of rust.

c. Must be designed with a view to stability, the assembly being such that when completely charged and ready for operation inclination at an angle of 30 degrees with the vertical will not result in upsetting.

d. Must be fitted with not more than one single or multiple burner, and the total rated gas consumption must not be more than ¾ cubic foot per hour.

e. Must be automatically regulated and uniform in their action, producing gas only as immediate consumption demands, and be so designed that gas is generated without producing sufficient heat to cause yellow discoloration of the residuum (which will occur at about 500 degrees Fahr.) or abnormal pressure at any stage of the process when using carbide of any degree of fineness.

f. Must have no mechanical or spring relief valve, and must

be so designed as to prevent automatically the accumulation of excessive pressure when placed in any position or overturned.

g. No valves or pet-cocks opening into the room from gas-holder parts, the draining of which would allow an escape of gas, are permissible, and condensation from all parts of the apparatus must be automatically disposed of without the use of valves or mechanical working parts.

h. Gage glasses, the breakage of which would allow escape of gas, must not be used.

i. The use of mercury seals is prohibited.

Mercury has been found unreliable as a seal in acetylene apparatus.

j. Combustible oils must not be used in connection with the apparatus.

k. Water seals, the breakage of which would allow escape of gas into the room, are prohibited.

l. Every lamp must be provided with a permanent marking stating plainly the amount of carbide necessary for a single full charge, the manufacturer's name and the name of the lamp.

31. *Generating Chambers.* a. Must be so designed that generation will take place under conditions similar to those which obtain in the best generator practice.

b. Must afford ample room for the residuum without containing unnecessary air spaces.

c. Must be so designed that the residuum will not clog or affect the working of the device, and be so arranged that the residuum can conveniently be handled and removed.

32. *Carbide Receptacles.* a. Must have sufficient carbide capacity to supply continuously the burner for which the lamp is rated during a lighting period of not less than six hours.

In determining charges the yield of gas from the various grades of carbide must be estimated as follows:

From the 3½ in. x 2 in. grade	4½	cubic feet	per lb.
" " 2 " x ½ " "	4½	"	"
" " 1¼ " x ⅜ " "	4¼	"	"
" " ¼ " x 1-12 " "	4	"	"
" " "Electrolite"	4	"	"

These figures are specified in order that a reasonable allowance may be insured for depreciation after the initial opening of the package.

The gas consumption of burners must be estimated at 50 per cent in excess of the rating.

b. Must be arranged so that the carbide holders or charges may be easily and entirely removed in case of necessity.

c. May in no case have capacity for more than two pounds of carbide.

33. *Water Supply.* Must be similar in quantity to the allowance made in the best generator practice.

34. *Pressures.* a. The working pressure at the burner must not vary more than ten-tenths (one) inch water column under all conditions of carbide charge and feed.

b. Lamps requiring pressure regulators or reducers must be so constructed as to withstand without injury a pressure equivalent to four times the maximum pressure obtainable in normal operation.

35. *Pressure Regulators.* a. Must conform to the rules for the construction of other acetylene apparatus so far as they apply, and must not be subject to sticking or clogging.

b. Must be capable of maintaining a uniform pressure, not varying more than four-tenths inch water column, at any load within their rating.

36. *Purifiers.* Where installed, purifiers must conform to the general rules for the construction of other acetylene apparatus and allow the free passage of gas.

37. *Means of Control.* a. Must be such as to obviate accumulation of pressure within the lamp or appreciable discharge of gas after extinction of the light.

b. Must be so arranged as to necessitate tight closure of the carbide filling opening before the feed can be put into operation.

Prominent features in the above rules, to which attention is here called, are:

(1) Use of liquefied acetylene or gas generated therefrom is absolutely prohibited.

(2) Generators must be well made and of substantial material.

(3) Generators must effect a regular and complete liberation of gas.

(4) Machines must accomplish *cool generation.*

(5) Safety is as near absolute as possible.

(6) Air mixtures and undue pressures are eliminated.

ACETYLENE AND CARBIDE STORAGE FORMS.

National Board of Fire Underwriters, 135 William Street.

NEW YORK, June 30, 1908.

At a meeting of the Executive Committee, held on the 25th inst., the following revised forms were adopted on recommendation of the Committee on Clauses and Forms:

NATIONAL BOARD STANDARD FORM NO. 12.

PERMIT FOR THE USE OF ACETYLENE GAS.

(AS AMENDED JUNE, 1908.)

In consideration of the following warranties on the part of the assured permission is hereby granted to light the premises described in this policy with acetylene gas, using a...........Acetylene Gas Machine manufactured by............at............

WARRANTED.

1. That the charging of the generator and the handling of the calcium carbide shall be by daylight only, and that no artificial light or fire shall be permitted within ten feet of the generator unless separated therefrom by brick or other non-combustible walls.

2. That no additions to or changes in the installation shall be made without notice to and the written consent of this company endorsed hereon.

3. *That the use of liquid acetylene or gas generated therefrom on the premises described herein, shall be absolutely prohibited.*

4. That in all respects the installation shall be in accordance with the rules of the National Board of Fire Underwriters.

5. That not more than 100 lbs. of calcium carbide will be permitted upon the premises, the same to be kept in water-tight metal cans and to be stored in a dry, waterproof and well ventilated place.

Attached to and forming a part of Policy No....of........Insurance Company of.............

..........................Agents.

CAUTIONS.

1. A regular time, during daylight hours, should be set for attending to and charging the apparatus.

2. In charging generating chambers, clean all residuum carefully from the containers and remove it at once from the building. Separate the unexhausted carbide, if any remains, from the mass and return it to the container, adding new carbide as required. Be careful never to fill the container over the specified mark, as it is important to allow for the swelling of the carbide when it comes in contact with water. The proper action and economy of the machine are dependent on the arrangement and amount of carbide placed in the generator. Carefully guard against the escape of gas.

3. Whenever recharging with carbide, always replenish the water supply.

4. Never deposit residuum or exhausted material in the sewer pipes or near inflammable material.

5. Water tanks and water seals must always be kept filled with clean water.

6. Never install more than the equivalent of the number of half-foot burners for which the machine is rated.

7. *Never test the generator or piping for leaks with a flame, and never apply flame to an outlet from which the burner has been removed.*

8. *Never use a lighted match, lamp, candle, lantern or any open light near the machine.*

9. See that the entire installation is in accordance with the rules of the National Board of Fire Underwriters, a copy of which may be obtained of your insurance agent, and obtain a written guarantee from the party installing same that these rules are complied with.

NOTE.—The failure to observe the above cautions is as liable to endanger *life as property.*

NATIONAL BOARD STANDARD FORM NO. 13.

PERMIT FOR STORAGE OF CALCIUM CARBIDE CONTAINED IN APPROVED METAL CANS HOLDING NOT MORE THAN 100 LBS. EACH.

(AS AMENDED JUNE, 1908.)

In consideration of the following warranties on the part of the assured, permission is hereby granted to store not exceeding six hundred (600) pounds of calcium carbide on the premises described in this policy.

WARRANTED.

1. That all calcium carbide will be contained in water-tight metal cans having a capacity not exceeding 100 pounds each.

2. That all such metal cans of calcium carbide will be stored in a dry, waterproof and well ventilated place and not more than one can of calcium carbide shall be opened for use or sale at any one time.

Attached to and made a part of Policy No.......of the........ Insurance Company of........................

....................Agents.

PERMITTED MACHINES.

It is still further to be noted that the Consulting Engineers of the National Board of Fire Underwriters examine and test generators submitted to them, and publish a list, revised from time to time, of machines which they approve. Factory practices used by manufacturers are also inspected from time to time and the manufacturing concerns are placed under agreement to furnish for sale under the names of permitted machines only such apparatus as are similar in all respects to samples and specifications on file at the Underwriters' Laboratories. The examination is rigid and searching, and while it may be that an occasional inferior and unsafe machine may have received the approval of the underwriters, answer may be made that the system, nevertheless, now rules out machines which do not comply with the requirements.

The public has thus been spared the danger which would have arisen from the unrestricted use of a large percentage of machines which failed to pass the inspection. There were also a multitude of others, especially in the early days of the industry, which never reached the inspection board, presumably because the makers of these machines were conscious of their defects. It is still further true that of the numerous different kinds of generators approved by the board, many, probably a majority, perhaps nearly all, failed to receive the official approval on first appearance and only did secure it after making repeated improvements at the suggestion of the board. The work of the underwriters is directed in the line of the common good, and aims at minimizing danger of fire; they are worthy of full credence when they attest, as they do, that the permitted machines "are safeguarded to as great an extent as it is possible to safeguard appliances of this character." The writer ought not to fail to include in his advice to persons who contemplate introducing acetylene illumination that they *purchase only such machines as continue to bear the approval of the underwriters.* There are plenty of such.

The latest list of such generators as the board will permit to be installed within insured buildings, issued under date of July, 1908, is here given. It is worthy of notice that out of about two hundred and fifty machines which passed the examination during the ten years from 1897 to 1908, inclusive, less than sixty remain on the latest list as now complying with the advanced requirements of the underwriters.

Attention is also called to the "Cautions" appended at the end of the list.

PERMITTED GENERATORS.

List of acetylene gas machines which have been examined under the rules and requirements of the National Board of Fire Underwriters by its Committee of Consulting Engineers, and which may be permitted for the uses specified when installed according to requirements, if the company insuring the property consents to the same in writing on its policy.

The following named acetylene generators made by the persons and firms whose names are given have been examined and tested under the direction of this committee, the factory practices used by these concerns are inspected from time to time and the manufacturers are under agreement to furnish for sale under these names only devices which are similar in all respects to samples and specifications on file at the Underwriters' Laboratories.

They are safeguarded to as great an extent as it is possible to safeguard appliances of this character.

Be sure that the trade-name, and the name of the manufacturer, as given in the list, are marked on the machine, and that the generator purchased is of the type manufactured since the date of the examination and test at the laboratories.

CLASS A.—STATIONARY APPARATUS FOR ISOLATED INSTALLATIONS.

Acetylene Report No.	Name of Machine.	Manufactured by	Manufactured at	Date of Original Exam.	Date of Last Re-exam.
198	Abner Giant	The Krein Mfg. Co	Wapakoneta, Ohio	July 1900	Nov. 1908
260	Abner, Jr.	The Krein Mfg. Co	Wapakoneta, Ohio	March 1901	June 1907
372	A Free Light	Martin Bros	Maryville, Mo	April 1905	June 1906

Acetylene Report No.	Name of Machine.	Manufactured by	Manufactured at	Date of Original Exam.		Date of Last Re-exam.	
261	Aurora	G. R. Lynch Mfg. Co.	Chattanooga, Tenn.	July	1901	May	1908
343	Ben Hur	The Johnson Acetylene Co.	Crawfordsville, Ind.	Jan.	1902	April	1908
438	Brauer	Brauer Acetylene Lighting Co.	Marshalltown, Ia.	Jan.	1907	June	1908
40	Buckeye	National Light and Heating Co.	Freeport, Ohio	April	1898	May	1908
419	Bunnell	Robinson & Love	Sycamore, Ill.	March	1906	March	1907
373	Carter	Niagara Falls Acetylene Gas Machine Co.	Niagara Falls, N. Y.	Jan.	1903	Feb.	1906
216	C. K. Sober	C. K. Sober	Lewisburg, Pa.	Nov.	1899	June	1906
121	Cleveland	Federal-Harris Mfg. Co.	Cleveland, Ohio	Jan.	1899	May	1908
346	Colt, Model N	J. B. Colt Co.	New York, N. Y.	April	1902	June	1908
420	Colt, Model O	J. B. Colt Co.	New York, N. Y.	Nov.	1905	June	1908
434	Comet	Comet Acetylene Apparatus Co.	Alexandria, Minn.	Aug.	1906	Aug.	1907
211	Condon's	U. S. Acetylene Co.	Philadelphia, Pa.	June	1907	....	
337	Davis, Model B	Davis Acetylene Co.	Elkhart, Ind.	Dec.	1901	Jan.	1907
383	Davis, Model C	Davis Acetylene Co.	Elkhart, Ind.	April	1903	Jan.	1907
36	Daylight	Daylight Acetylene Gas Co.	Louisville, Ky.	March	1898	April	1908
450	Daytonia	St. Clair Mfg. Co.	Dayton, Ohio	Sept.	1907	....	
1	Eagle	Eagle Generator Co., St. Louis, Mo.	Chicago, Ill.	June	1897	March	1908
406	Federal	Federal Acetylene Co.	Chicago, Ill.	Aug.	1905	March	1907
265	Gate City	Visser & Brown	Keokuk, Ia.	April	1901	April	1908
391	General	Commercial Acetylene Co.	New York, N. Y.	Sept.	1903	June	1908
369	Grand Rapids	G. F. Owen	Grand Rapids, Mich.	Jan.	1903	Nov.	1907
418	Hays	J. T. Hays & Son	Emmitsburg, Md.	Jan.	1907	....	
448	Hercules	G. R. Lynch Mfg. Co.	Chattanooga, Tenn.	Jan.	1907	May	1908
365	Higgs	Davis Acetylene Co., for Home Carbide Co., Chicago, Ill.	Elkhart, Ind.	Jan.	1903	July	1907
453	Hinkle's Measured Feed	Globe Cornice Works	Decatur, Ill.	Aug.	1907	....	
440	Holm	Springfield Acetylene Generator Co.	Springfield, Ohio	Dec.	1906	Dec.	1907
273	Hunter's Carbide Feed	Hunter's Acetylene Gas Co.	Elkhart, Ind.	Nov.	1900	Jan.	1907
240	Ideal Epworth, Type D	Ideal Epworth Acetylene Co.	Waterloo, Ia.	Sept.	1900	June	1908
240	Ideal Epworth, Type D	Ideal Epworth Acetylene Co.	Johnstown, Pa.	March	1907	....	
403	Illinois	Monmouth Acetylene-Electric Mfg. Co.	Monmouth, Ill.	Nov.	1905	Sept.	1907

338	King	C. A. King	Norwich, N. Y.	Oct.	1902	Jan.	1908
71	Laun	Laun Bros	Chicago, Ill.	Aug.	1898	Feb.	1908
464	Model	National Welding & Mfg. Co.	Buffalo, N. Y.	May	1908	......	
253	Monarch Carbide Feed	Monarch Acetylene Gas Co.	Omaha, Neb.	April	1901	May	1908
253	Monarch Carbide Feed	Monarch Acetylene Gas Co.	Buffalo, N. Y.	July	1904	May	1908
410	Night Commander	United States Standard Co.	Voorheesville, N. Y.	Aug.	1905	Nov.	1907
282	Orion	Matteson Acetylene Gas Generator Co., successors to Boonville Acetylene Generator Co., Boonville, N.Y.	Pulaski, N. Y.	Oct.	1904	Jan.	1908
340	Phelps Carbide Feed	The New England Mfg. Co., Chicago.	Elkhart, Ind.	April	1903	Jan.	1907
454	Phelps Measured Feed	The New England Mfg. Co., Chicago.	Elkhart, Ind.	July	1907	......	
404	Pilot, Model B	The Acetylene Apparatus Mfg. Co.	Chicago, Ill.	Oct.	1904	Nov.	1907
386	Pilot, Model C	The Acetylene Apparatus Mfg. Co.	Chicago, Ill.	Feb.	1903	Nov.	1907
428	Pilot, Model D	The Acetylene Apparatus Mfg. Co.	Chicago, Ill.	May	1905	Nov.	1907
381	Radiant	The Rush Acetylene Generator Co.	Canandaigua, N. Y.	Feb.	1903	April	1906
388	Reliance	The Gem City Acetylene Generator Co.	Dayton, Ohio	Dec.	1904	Oct.	1907
460	Success	Springfield Heating & Ventilating Co., for S. T. Russell	Springfield, Ohio	Feb.	1908	......	
354	Sunlight Omega	Sunlight Gas Machine Co.	New York, N. Y.	July	1902	June	1908
347	Victoria Model B	Victoria Mfg. Co.	Auburn, Me.	Feb.	1902	Sept.	1906
430	Aceterator	General Light, Power & Pump Co.	Philadelphia, Pa.	Sept.	1906	Jan.	1908

This generator is considered suitable for use only when installed in a masonry pit, with the generating water below the frost line, located outside of and at least 30 feet removed from buildings.

CLASS B.—STATIONARY APPARATUS FOR CENTRAL STATION SERVICE.

Acetylene Report No.	Name of Machine.	Manufactured by	Manufactured at	Date of Original Exam.		Date of Last Re-exam.	
239	Williamson	Acetylene Apparatus Mfg Co.	Chicago, Ill.	April	1902	Nov.	1907

CLASS D.—PORTABLE TABLE LAMPS.

Acetylene Report No.	Name of Lamp.	Manufactured by	Manufactured at	Date of Original Exam.		Date of Last Re-exam.	
429	Beck-Iden	Acetylene Lamp Co.	New York, N. Y.	Sept.	1905	June	1908

4. Whenever recharging with carbide, always replenish the water supply.

5. Never deposit residuum or exhausted material from water feed machines in sewer pipes or near inflammable material.

6. Water tanks and water seals must always be kept filled with clean water.

7. Never install more than the equivalent of the number of half-foot burners for which the machine is rated.

8. *Never test the generator or piping for leaks with a flame, and never apply flame to an outlet from which the burner has been removed.*

9. *Never use a lighted match, lamp, candle, lantern or any open light near the machine.*

10. See that the entire installation is in accordance with the rules of the National Board of Fire Underwriters, a copy of which will be furnished by your insurance agent, and obtain from your contractor a written guarantee that these rules are complied with.

NOTE.—Failure to observe the above cautions is as liable to endanger *life* as *property*.

ACETYLENE ILLUMINATION A MEASURE OF SAFETY.

It is the business of the fire insurance underwriter to take cognizance, in his conservative way, of every new process, or industry, which affects the fire hazard in any way and to adjust his rulings to changes of condition. If these be in the direction of greater safety, he will encourage them; if of increased hazard, he will do the reverse.

No innovation can come much nearer to the heart of the underwriter than a new method of illumination. Candles came into general use long before the underwriters were organized and, however great the hazard they cause, they are hardly under the jurisdiction of the underwriter. Acetylene, on the other hand, a development of the last dozen years, and now so far advanced that its brilliant light encircles the globe, and that its use in the United States alone exceeds twenty billion candle power hours per annum, has grown into popular favor under the complete control of the National Board. The advance of the new illuminant has been permitted only after its competent committee of consulting engineers have rigidly held to every detail of requirement, which they, as representatives of the Board, have themselves established.

Acetylene does cause a certain measure of fire hazard, it is true. But so also does the man who throws his lighted cigarette into the waste basket, or the lady of the house who cleans her gloves or her canary bird with gasoline, and so did the servant girl who lighted the fire with benzine, and about whom the local paper unfeelingly remarked that "nothing had *benzine* of her since!" So likewise do lighting hazards arise from the use of liquid hydrocarbons in any form, of matches and candles, and of electric lighting. But when we recall that acetylene is used in so many homes, largely in country places, that when a million burners are lighted, as occurs every evening, replacing as they do the candle with all its dangers as well as gasoline and kerosene which may be spilled or used to encourage the kitchen fire, can there be any denial of the truth of the assertion that acetylene in such cases diminishes the hazard?

It seems to us that we may assert with positiveness that the introduction of acetylene does produce a distinctly lower fire risk than existed in the same localities before. If not, wherein lies the danger? Certainly not in the carbide, whose properties we have described fully elsewhere, and which is transported at the lowest freight rates, long experience with large shipments having convinced the transportation companies that it can be handled literally without hazard. It certainly does not reside in the water, the only other substance which is required for the generation of the gas; it is therefore not *behind* the generator so to speak. Now, to the *front* of the generator we find only the pipe system, laid in accordance with underwriters' rules, fitted with a minimum of leakage, because this gas being more concentrated than other illuminating gas, every owner looks carefully after his pipes as a measure of the commonest economy. Then come the burners, about which, it has been pointed out that if a full sized burner should by any chance be left open in a sleeping room, and should the occupant fail to be roused by the odor, or be unable to escape from the room, he would die of starvation before the air would become irrespirable. Compare this with the inevitable result of leaving the coal gas burner open in a small sleeping room for a few hours; and then recall that even if a large leak should occur, from a broken fitting or a disconnected pipe, the ordinary house generator could only deliver the result of a single charge, probably not over a hundred cubic feet, whereas a similar leak in a city supply pipe has behind it the entire contents of the city gas holder!

Now if the hazard is neither to the front nor to the rear of the generator, we have only the generator itself yet to inspect. The acetylene generator is the only illuminating contrivance which has enjoyed, from its inception, the guidance of the underwriters. The industry has paid the tax and heartily coöperated with the underwriters in their efforts to attain their ideal. Safety devices lock and interlock the different parts exactly as prescribed by the underwriters. Their trained guidance has determined the thickness of the metal employed, the feeding devices, escape valves, means for preventing over-feeding, for stopping the action of the machine should any detail be out of order, etc., etc. Their primary idea has been to produce an ideal machine, so nearly as possible perfect for inside installation, and it may be truthfully said *that they have developed a generator which assures a smaller risk than the illuminants it replaces.*

RESOLUTION ON ACETYLENE

PASSED BY THE NATIONAL BOARD OF ENGINEERS AND REFERRED BY THEM TO THE NATIONAL BOARD COMMITTEE ON LIGHTING, HEATING AND PATENTS, BY WHOM IT WAS SENT TO THE EXECUTIVE COMMITTEE OF THE NATIONAL BOARD OF FIRE UNDERWRITERS, WHO, ON JANUARY 30, 1908, MODIFIED THE RULES IN ACCORDANCE WITH THE SUGGESTIONS OF THE ENGINEERS.

In view of the fact that the number of acetylene generators installed inside of buildings has very largely increased in the past few years, while at the same time our records seem to show that the number of fires ascribed to such installations has, if anything, decreased, thus showing that the rules of the National Board covering the construction in installation of acetylene apparatus has apparently safeguarded the hazard to a very great extent; and

In view of the fact that the introduction of this method of illumination generally replaces some methods which entail an equal if not greater hazard; and

In view of the fact that outside installation is frequently impracticable on account of climatic conditions;

THEREFORE, we are of the opinion that the time has arrived when the National Board rule as to outside location may properly be modified so that, while it will recommend the outside location as being the ideal location, yet it will permit, under proper restrictions, the location of the generator inside of the insured building, and we therefore recommend that Rule No. 2, Section A, of the National Board rules be amended to read as follows:

Generators, especially in closely built up districts, should preferably be placed outside of insured buildings in generator houses constructed and located in compliance with Rule No. 9,

COST OF ACETYLENE ILLUMINATION.

The cost of acetylene illumination is necessarily made up of two elements: (1) the first cost of the generator and its installation, and (2) the current expense for carbide.

The first item, cost of generator and its installation, is of necessity a variable quantity, and no attempt would better be made here to fix upon a figure. The size of the machine must naturally be adapted to the space which is to be lighted but, in general, it would be well to purchase abundantly large for present needs, as these may become increased and new needs may develop. If the buildings have already been piped for illuminating gas, the old piping and fixtures will do equally well if not better; the burners alone will need to be changed. If new piping is required, only about half the size requisite for city gas is usually employed for acetylene, and the cost of this is a matter of local conditions. Nor need the matter of placing a pipe system even in a completed and finely finished residence be regarded as objectionable. With a little care it can be thoroughly concealed beneath the floors and within the walls.

The present cost of carbide may be stated at seventy dollars per ton, or in small quantities three and three-fourths cents per pound. On the Pacific coast, owing to high freight rates the price is ninety-five dollars per ton, or five cents per pound. Since a pound of carbide yields, as a conservative estimate, four and a half cubic feet of gas, a safe and reasonably close approximation is made when the cost of acetylene is stated as distinctly less than one cent per cubic foot, aside from the first cost of the installation. And when it is considered that the commonest acetylene burner in use (one-half foot) consumes only one-tenth as much gas per hour as the usual coal gas burner (five-foot), acetylene is found competing with coal gas at ninety cents per thousand, a lower price for the latter commodity than is often realized. Still further, if the greater illuminating power of acetylene be considered, it may be stated with little likelihood of challenge that, *light for light,* acetylene is cheaper than coal gas. The writer has seen a large number of letters from householders, storekeepers and others using acetylene gas, who declare without reserve that their present light is not only vastly better, but also much cheaper than their former coal oil illumination. Again, as the acetylene flame is usually burned free, without shade or chimney, there may be at once eliminated from all estimates any allowance for expense of broken globes, chimneys, shades or incandescent mantles, an item of no inconsiderable importance which is often underestimated in making calculations of prospective expense. Another matter of importance is that all the light produced is utilized without diminution of its intensity by opaque globes, shades or chimneys, for it is to be remembered that even clear glass absorbs a perceptible percentage of light which is never utilized by those who derive their artificial light from flames surrounded by glass or porcelain.

It is true that many an acetylene trade catalogue presents figures so much more favorable to acetylene than the above that, if credence were accorded them, one could but wonder why acetylene does not at once expel coal gas and electricity from use. The sophistry and bombast of such irresponsible people is to be carefully avoided, lest one be led astray. On the other hand, the devotees of coal gas, the journals which represent the capital invested in that industry, may very likely display calculations at times which would indicate that acetylene must promptly and irrevocably fail. As between the two, the strides with which acetylene has bounded into popularity must serve to direct our judgment, and while the contention can hardly be that acetylene is fitted by its nature and its present cost, to compete with coal gas in the cities and towns, the statements given above may be reasonably relied upon as showing the fitness of this illuminant for inexpensive and beautiful lighting of isolated factories, of churches, hotels, schools, institutions, stores, small country villages, and particularly rural homes. And the figures given in this chapter on the expense of acetylene illumination may be treated as reasonably reliable, showing the results of the best experience at the present time, assuming always a well made, safe, reliable generator of responsible manufacture.

BURNERS.

It is not practicable to utilize acetylene in the usual burners employed for other gas. The burners must pass less gas and be especially constructed to meet the demands made upon them. Much ingenuity has been expended in producing good acetylene burners, with the result that there are on the market to-day a large number of different types which serve their purpose admirably.

The first and most serious difficulty was in finding a burner in which acetylene could be consumed without the nuisance of smoking and clogging. This was because the burner of ordinary form becomes too hot; the gas polymerizes as was explained under the properties of acetylene; the condensation products penetrate the material of the burner and, charring, clog the passages with soot. The problem of finding a burner which should not exhibit this objectionable character was solved by the introduction of burners in which two small jets of acetylene are discharged from two different points in the burner, so directed that they impinge upon each other at a slight distance from the burner, forming a flat flame which stands at a right angle to the plane in which the jets lie. These jets also carry air with them, taken in at the burner by the aid of a device which is similar in principle to a Bunsen burner on a small scale. The injected air is merely sufficient to bring about a complete and smokeless combustion, and the flame itself is so remote from the burner that the latter scarcely becomes heated.

In some of the earlier forms of branched burners in which metal arms were employed, it was found that even the slight heating to which they are subjected would, in time, warp the tube so that they would cease after a while to work "true," and would begin to smoke when used but a few hundred hours. This was obviated by a modified construction of the metal burner or by substituting steatite, which is a poor conductor of heat, for the metal in the branching portion, and thus, in different ways, burners are now available which do not either smoke or clog after some thousand hours of use. A burner in a buoy under the inspection of the Canadian Government was recently found in perfect condition, and giving full light, after fifteen months' continuous service without attention.

More recently burners have been invented which produce the flat flame without employing a branched construction. It is claimed for such that, having no jets to align, they can not get out of alignment. This is certainly an advance in burner making if it shall succeed in producing the other necessary qualities of a perfect burner; but, after all, failure of alignment is for the most part a thing of the past, and there are plenty of good burners to-day which fulfill all requirements to an admirable degree, as may be seen by a glance at the advertising pages of the Acetylene Journal. Much ingenuity has also been expended in the production of satisfactory turn-down burners, and this has not been wholly without success. Burners are made in sizes ranging from one-eighth to one and one-fourth foot, and should be marked on the pillar. Competition in burner manufacture has been such that this very necessary adjunct to acetylene illumination has been most perfectly cared for, and no improper or inefficient or uneconomical burners can to-day find any sale. The purchaser cannot go far astray in this matter.

It may be remarked here that the quality of the gas has much to do with the excellent operation of the burner. A gas which has been overheated in the generator in disregard of all precautions concerning cool generation, a matter under the absolute control of any one selecting a generator, will most surely carry with it products other than pure acetylene, which will cause it to behave badly even in the most approved and perfect burners.

Again, it is false economy to turn down the ordinary form of acetylene burner, as this causes the air injection to refuse to work, brings the flame nearer the burner, heating the latter and thus defeating the very object of its careful construction and causing it to become clogged by the deposit of solid matter in the minute passages in the burner, which once accomplished is difficult if not impossible to remedy. The flame of all acetylene burners is small, but there is no dark, non-illuminating portion common with other gas. We are likely to underestimate its lighting power because of its small size, which, however, is more than offset by its great intensity. All good burners produce a delightfully steady, quiet and agreeable flame, free from flickering.

If anything can be judged from the discussions in the various journals on the principles underlying the production of light by processes of combustion, we might reasonably conclude that the ways in which acetylene is consumed at the present time by no means exhaust the possibilities of our illuminant. But this is no place for a consideration of the various problems which arise in burner construction, and all of the novelties in acetylene combustion will be deferred until another time.

FREEZING.

As water is necessarily used to produce acetylene, and as it is also usually employed for seals, as well as for washing the gas and for the operation of the holder, it follows that acetylene generators must be placed where they can not freeze. On the other hand, there should be no fire near such a machine. The proper arrangement is to install a generator in a protected space where the temperature never falls below freezing, or supply the generator room with a small radiator or coil fed with hot water or steam. If installed in a separate generator building, such building may be made practically frost-proof by providing two or three dead air spaces in the walls, giving it double doors and windows, and taking such other precautions as may suggest themselves. No free fire can ever be permitted under any circumstances, but a coil of hot water or steam pipe, when it can be inexpensively provided is an excellent safeguard. If installed within the building lighted, similar precautions may be properly taken, and whatever the method of securing the result, it is absolutely essential that the generator be so placed that no freezing can take place.

DISPOSITION OF RESIDUUM.

The liquid and semi-liquid residue may be passed into the sewer, provided there is sufficient fall and no scarcity of flushing water, and allowed to accompany the sewage, upon which it can exert no action or influence, other than that of a partial disinfectant. The use of the agitator with which most generators are provided, facilitates this removal of refuse. But if the drain is too flat, the sludge will settle in a compact mass, which eventually stops the pipe. In many cases, therefore, it is far preferable to allow the sludge to flow through a separate drain with plenty of fall and permit it to accumulate where it may subsequently be utilized if in sufficient quantity. If care is taken in the construction of a sludge pit, so that the residuum will be drained as well as possible, and yet not placed so deep as to be difficult to load it into carts for removal, this material may even become of some pecuniary value.

The chief uses of the sludge, frequently called acetylene lime, are for mixing mortar, for whitewashing fences, cattle pens, fruit trees, etc., for making paths, and for fertilizing, with some occasional application as an insecticide and disinfectant. Mortar made from it is reported to bind quickly and hard; there is no reason why it should not be at least of equal value with mortar made from ordinary slaked lime. It may be added that in any of the multitudinous uses to which ordinary lime whitewash is applied, a whitewash made from carbide residuum answers equally well. The disposal of residuum depends on local conditions; manufacturers of generators usually possess information in the matter which enables them to advise, under specific circumstances, so that no general statement can finally solve the problem.

In view of the many practical uses to which acetylene lime has been successfully applied, and particularly because of its usefulness as an application to the soil, it may not be out of place to submit the chemical analysis of carbide residuum. The following figures show the analysis of three specimens of residue taken at remote times and places:

	I. Per cent.	II. Per cent.	III. Per cent.
Sand (silica)	1.24	1.10	0.97
Carbon (coke or coal)	2.08	3.95	2.14
Oxide of iron and alumina	3.11	2.90	2.30
Lime	62.50	63.65	66.10
Water and carbonic acid	31.07	28.40	28.49
Total	100.00	100.00	100.00

I. is an analysis of sludge from a German carbide; II. is an analysis by Professor E. H. Jenkins, of New Haven, Conn.; III. is an analysis made by the writer at The Pennsylvania State College. In order to present an authoritative statement of the fitness of such residues for application to the soil, the above analyses were submitted to Dr. William Frear, Chemist to the Pennsylvania Department of Agriculture, with a request for a statement, regarding the use of residuum of such character for agricultural purposes. The following response was received:

State College, Pa., Dec. 28, 1907.

Prof. G. G. Pond, State College, Penn.

Dear Sir: The three analyses submitted by you of residuum from acetylene machines show it to consist principally of partially carbonated slaked lime. Its composition is very nearly that of the lime slaked in the field for agricultural use. I know of no reason why this acetylene residue is not well fitted for the same agricultural use. On the basis of the analyses submitted, one ton of the residuum would be equivalent to approximately fifty bushels (forty pounds to the bushel) of the slaked lime above mentioned, or to twenty bushels of the best stone lime (seventy-two pounds to the bushel).

Very respectfully,

(Signed) WILLIAM FREAR, Chemist.

PURIFICATION.

The question of the purification of acetylene has been vigorously agitated at various times, especially in Germany and France. Some excellent suggestions have been made, and various chemical purifiers proposed, some of which are of undoubted merit.

It should be understood, as has been mentioned before, that the first and most important purification to which acetylene is subjected is the washing which it receives by bubbling through the water as it must do in all forms of generators; then it is always "filtered" through felt, a purifying treatment which has been found invariably necessary in order to remove solid particles or dust which float in the crude gas, chiefly lime dust.

Purification, therefore, in the sense in which the word is used in this chapter, means chemical purification, the aim of which is to remove those minute traces of ammonia, sulphureted hydrogen, and particularly phosphorus-containing compounds which escape the washing apparatus. Throughout the discussion we must bear in mind that these impurities can arise only from impurities in the carbide from which the gas is made, and that American carbide, as it is produced to-day, is of such high grade that the question of purification is by no means a vital one. The removal of the last traces of accompanying impurities in commercial acetylene is not a necessity, though it may be a thing to be desired. Most of the hundred and fifty thousand homes lighted by acetylene in this country to-day, employ no method of purification other than the water washing and filtering mentioned above and to which the gas is submitted in every good generator, yet no harm whatever results from this omission. Nearly all of the American manufacturers believe that chemical purification is unnecessary. But if anyone desires to free his gas a little more perfectly, if possible, from the last traces of accompanying impurity, it can be most easily accomplished by means of readily available purifying apparatus which can be attached to any machine or inserted between the machine and the distributing pipes with little trouble and at trifling cost. Four purifying mixtures which have been proposed, and used with more or less success, are:

1. *Heratol.* A solution of chromic acid or sulphuric acid absorbed in porous earth.
2. *Acagine.* A mixture of bleaching powder with fifteen per cent of lead chromate.
3. *Puratylene.* A mixture of bleaching powder and hydroxide of lime made very porous, and containing eighteen to twenty per cent. of active chlorine.
4. *Frankoline.* A mixture of cuprous and ferric chlorides dissolved in strong hydrochloric acid absorbed in infusorial earth.

A discussion of the relative merits of the chemical purifiers would lead too far into the realm of chemistry for the limits of a popular bulletin. The writer still maintains the belief that, so long as the present high quality of American carbide is maintained, purification may just as well be omitted in small installations.

MUNICIPAL LIGHTING.

The usefulness of acetylene is not necessarily limited to small installations adapted to single small buildings. Larger establishments such as factories, groups of summer hotels, army posts, schools or colleges, public institutions, etc., particularly when somewhat isolated in location so as to be beyond the reach of city gas supplies, may often be very beautifully and economically illuminated with acetylene. Plants of sufficient size to serve for the illumination of villages and small towns, ranging from five hundred to five thousand inhabitants, are becoming more and more numerous every year. There are several hundred such acetylene towns in the United States and similar progress has been made in Germany, France, Austria and other countries. A few towns with central acetylene plants have been reported as giving up this form of illumination after a season of trial and some such have been exploited in the public press, to the detriment of the industry. But such instances have been, almost exclusively places where, by reason of imperfect installations or unskillful pipe laying, the outfits have been of too cheap an order to survive, and no case has come to the attention of the writer in which a really excellent and well equipped acetylene lighting station has been discarded in favor of any other form of public illumination. Prominent among the features which require to be especially looked after in successful town-lighting are:

1. Solid and substantially constructed generators of approved type.

2. Sufficient gas storing equipment so that there need be no question as to abundance of gas on hand.

3. Perfect pipe fitting so as to minimize leakage.

Generators for town lighting are in many cases not of the automatic type; that is the generation of gas is not controlled by the supply in the bell as in the small household generator, but they are fed by hand, or better by a small motor so that a slow and steady feed is assured. It is customary to generate enough gas at a time to last at least over a single night and often for several nights. To this end ample storage is desirable, so that the gas may be completely cooled as well as to avoid the possibility of the gas supply failing when lights are burning, and being turned on again while the burners are still open; very perfect pipe laying is essential in order that there shall be no unnecessary leakage since acetylene is a very concentrated illuminant, possessing large candle power in small bulk and hence any considerable loss of gas by leakage means decided reduction in profits.

Consumers generally express themselves as greatly pleased with the quality of the light, the excellence of the service and the cheapness of the illumination. Street lamps and other forms of public illumination are provided as well as the house-to-house service, and satisfactory meters are on the market, especially adapted to the requirements of acetylene. This form of acetylene apparatus is also carefully safeguarded by the underwriters; a central station is erected in a manner similar to that used for coal gas lighting, with this important difference that it is much smaller and consequently less expensive. All the cumbersome, expensive and care-demanding equipment including retorts, scrubbers, etc., essential for coal gas are entirely unnecessary. A small, compact generating system is all that is required and gas is produced in precisely the same manner as in a private installation.

ACETYLENE TABLE LAMPS.

Our illuminant is unique for its great flexibility of application. We have considered its perfect adaptability to the requirements of a town and its admirable fitness for being selected as the illuminating agent for a house of any dimensions. But we shall fall far short of exhausting this singular flexibility if we do not introduce our readers to the portable lamps which may perhaps have attracted their attention in the underwriters' regulations.

These admirable contrivances, no less truly than larger domestic equipments, are easily and cheaply available at the present day. They are little automatic generators in themselves, adjusted to feed a single burner, and may easily be carried from room to room. They are charged by filling the base with about a gallon of water, and placing in the urn a proper quantity of especially prepared carbide, of a size which roughly resembles granulated sugar. They work admirably, furnishing a smooth, unwavering flame. Daily filling is desirable, though not necessary, but it is more easily accomplished, and cleaner work than the daily attention required by an oil lamp of the same size. The cost of operating the portable lamp is a trifle greater than that of other forms of acetylene lighting, because of the somewhat increased cost of the prepared carbide, known as *Electrolite,* which is packed expressly for use in these lamps, but one cent per hour is the allowance of cost which the writer estimates will supply his study table with this restful convenience. The portable acetylene lamp is a distinct advance in civilization; it is fully as efficent as that grotesque Vesuvian monstrosity, known as the piano lamp, and greatly to be preferred on grounds of care, convenience and safety. It also possesses the enormous advantage that it contains no inflammable material to spill and cause disastrous conflagration; it may be overturned while burning, without any likelihood of harm other than the possible breakage of the fragile parts, or indenting of the metal.

OTHER PORTABLE DEVICES.

Nor does the table lamp exhaust the smaller and portable apparatus which make use of calcium carbide and acetylene. There are hundreds of thousands of lamps in daily use, employing acetylene as a means of illumination for bicycles, carriages, motor cars, etc., with which everybody is familiar and which call for no other comment here except to say that acetylene has proved its usefulness for this kind of illumination, in which we may include also vendors' torches, torches for lamp-lighters, searchlights, miners' lamps, which in Germany have become a staple article, and in this country are already selling in great quantities, automobile mail cars and delivery wagons, ploughing, threshing and other farming operations at night, headlights for traction engines and steam shovels for night work, lanterns, apparatus for projection of pictures either on the small or on the large scale, mechanics' torches and so on.

COOKING BY ACETYLENE.

A few years ago cooking by means of acetylene was thought to be entirely out of the question, by reason of expense and impracticability of securing suitable burners. It is safe to assert today that the burner problem has been very perfectly solved and such a measure of economy in heat production brought about that cooking by acetylene is both entirely practical and reasonably inexpensive. Excellent blue flame burners are now readily obtainable anywhere, and complete cooking ranges, equipped with a variety of such burners, are to be found in the stock of every up-to-date dealer. Cookers, hot-plates, chafing dish heaters, in fact stoves of every sort which have long been employed for coal gas are now adapted to acetylene, and are giving satisfaction in thousands of homes.

Other accessories for the household are instantaneous water heaters, flame spreaders for heating sad irons, radiators for room heating, fire places, and grates. It is literally true, and no longer the dream of the visionary enthusiast that, with the use of acetylene for lighting, cooking and heating, the out-of-town resident is placed in a position to enjoy equal conveniences with the resident of the city, if not even superior ones.

DISSOLVED ACETYLENE.

The remarkable application of acetylene, which will be described under this caption, is one which has extended the usefulness of the gas to a very enormous extent and lies at the basis of its extension in so many ways not known before that no discussion of acetylene is complete without a full consideration of dissolved acetylene. We have had occasion to mention, in an earlier chapter, the small degree of solubility in water which acetylene possesses. It is of consequence for us to know about it, for the excessive use of water might cause considerable loss of gas. But the solubility of the gas in water is too limited to be put to any practical use. Its solubility in other liquids is not of more interest to us except in a single liquid which, though familiar to the chemist, is not commonly known to the public as it meets no important uses in the arts or manufactures. *Acetone,* one of the products of the distillation of wood, and prepared also in other ways for the market, is a limpid, colorless, mobile, fragrant liquid, boiling at one hundred and thirty-three degrees Fahrenheit, inflammable and burning with a luminous flame, an excellent solvent for many resins and gums. It is four-fifths as heavy as water, is capable of mixing freely with water, alcohol or ether; it has a peculiar but not disagreeable odor and a biting taste not unlike peppermint; it is much employed in the manufacture of chloroform, iodoform and certain other medicinal preparations. This long known but rather unfamiliar fluid is a peculiarly excellent solvent for acetylene. Acetylene dissolves in it as freely as sugar does in water, and the solubility increases with the pressure applied. Acetone dissolves twenty-four times its own bulk of the gas at ordinary temperature and pressure, and for each additional atmosphere of pressure to which it is subjected a similar quantity will be dissolved, in other words the quantity dissolved increases directly with increase in pressure.

Now the compression of acetylene to an extent greater than two atmospheres, and particularly to the liquid condition which is reached at about forty atmospheres, at ordinary room temperature, has been shown to be accompanied with so much danger to endothermic explosion, that the practice has been given up. But if the gas be forced into this remarkable liquid, acetone, a mixture is secured which seems to be devoid of the dangerous quality of other forms of compressed acetylene, and hence available for safe transportation and use.

But at the same time that acetone dissolves acetylene, it increases its volume very perceptibly in so doing. When the pressure is released the acetylene escapes, and can be utilized at pleasure, but the acetone reassumes its smaller volume and thus produces a space in which "free" acetylene would of necessity exist, and create a source of danger. Of course this possibility of accident had to be removed before dissolved acetylene could be a commercial success.

This important problem was solved by first filling the receptacle with some porous solid material which should leave no space whatever to be subsequently occupied by "free" compressed acetylene. Various absorbent materials were employed for this purpose, chief among them being asbestos, wood charcoal, infusorial earth, etc., the aim being to secure a material which should possess a porosity of seventy-five to eighty per cent. and at the same time remain inert towards the acetone and acetylene which it is destined to absorb. In practice, the industry has settled upon asbestos, and by careful packing of cylinder or tank, making use of such a system of devices that there can be no possibility of any free space, either when the cylinder is freshly prepared, or by subsequent sagging down of the asbestos, this kind of apparatus has reached a very high degree of perfection.

Briefly the method of preparing dissolved acetylene is as follows: The gas is generated in the usual way, is carefully washed, purified and dried. Great care is exercised to keep the acetylene as free as possible from impurities, and to exclude so far as possible, all admixture of air. The compressers force the gas into large storage tanks or accumulators containing asbestos saturated wtih acetone as above described. The pressure is allowed to increase to one hundred and fifty pounds and the storage tanks are connected by means of a system of steel pipes to the cylinders, previously charged with porous material and acetone, for the final reception of the gas for transportation and use.

The measure of safety which is claimed for this storage system, as compared with other forms of compressed acetylene, resides in the degree of perfection with which the tanks are so completely filled with the absorbent material that there is no possibility of free, or undissolved, gas anywhere in the system, and secondly upon the fact that the solution of acetylene in acetone is inexplosive even under high pressures. Its freedom from likelihood of accident has been demonstrated to the entire satisfaction of many railroad officials on whose roads the system is employed.

Since the gas issues from the tanks, whenever released, at a high pressure, it is necessary to employ a pressure regulator to bring the pressure down to the correct extent for practical use. But such apparatus is easily secured, inexpensive and reliable. Dissolved acetylene possesses the advantage of being delivered at the burners cool, dry and at uniform pressure, the minute quantity of acetone vapor which accompanies it possessing no disadvantageous quality. It can be turned on and off at any moment without waste of gas, cannot freeze and the quantity of gas left in the tank is always known by a glance at the gauge.

The applications of dissolved acetylene are numerous, and increasing daily. For ordinary house lighting, it is entirely out of the question; acetylene can be provided for stationary purposes by generators already described so much more economically that dissolved acetylene finds no application. But where portability is a requirement, as on yachts and excursion barges, for vehicles of all sorts, motor cars, railway coaches, locomotive head-lights, etc., dissolved acetylene finds ready and welcome application. More particularly where accessibility is so difficult that the use of a generator is prohibited, as in buoys and lighthouses, some of which can only be approached at long intervals and in favorable weather, there is a demand for a form of illuminant which dissolved acetylene is most admirably adapted to fill.

ACETYLENE FOR LIGHTHOUSES, BUOYS AND SIGNALS.

The United States Lighthouse Board commenced experimenting with acetylene for beacons and other marine lights as early as 1896. The generators of that day were not highly perfected, and progress was retarded, but within the last five years much has been done in the way of utilizing this illuminant for the maintenance of lights for the guidance of mariners. Some thirty-five generators have been installed in Mobile Harbor and others in the Harbor of Galveston. Some of them have been in continuous service since early in 1902. The saving in cost of maintenance, as compared with the old oil system, is nearly fifty per cent. Other acetylene lights are maintained off Cape Henry, off Point Judith, at Sandy Hook and other points. Most, if not all, of these installations are carbide-feed generators; some of these receive attention every sixty days; others are self-sustaining for a much longer time, even up to six months; and pilots and masters of vessels speak in high terms of the excellence of the lights and particularly their fog-penetrating character.

The Canadian Government has, since 1901, been maintaining acetylene lights of various kinds in Canadian waters, and such lights are much in favor with the Lighthouse Service officials at Ottawa, who have expended a large percentage of their appropriation upon the installation of Willson acetylene buoys, most of which are of the low pressure type. Such a buoy can carry from nine to ten thousand feet of gas, in the form of carbide; can be charged at the opening of navigation and requires no attention, so far as supply is concerned, until navigation closes. There is no artificial illuminant known, except acetylene, which can be left to burn unattended, and without replenishing the supply of fuel, for anything like so long a period. Such rapid progress has been made by acetylene in replacing other illuminants in the lighthouse service in various parts of the world, the coast of France, the Isle of Shetland, Scotland, Alaska, etc., that it is unnecessary to further accumulate data to show that acetylene already occupies an important place as an aid to navigation. It is quite obvious that the very qualities which cause this gas to occupy the advanced position which it does, as a domestic and municipal illuminant, namely efficiency, economy, and small care in operation, are the very ones which influence governmental use of the same light in aiding navigation.

RAILROAD USES OF ACETYLENE.

The advantages of acetylene as an illuminant have from the first appealed to the railroad officials of all countries. Aside from the lighting of small stations, its use has been chiefly for two purposes, locomotive headlights and the illumination of coaches. As an enricher of coal gas and oil gas, acetylene made some advances long ago, but it was not destined to achieve its great ascendency in this way. Portable generators on coaches have also been tried out to a considerable extent, and not largely adopted, but with the advent of dissolved acetylene, the problem seems to be efficiently solved. Dissolved acetylene is now used by hundreds of railroads for locomotive head-lights, and while it is not possible to secure so high a candle power as electricity, the light is far more efficient and satisfactory, at one-fifth the cost of an electrical arc headlight. Many leading railroads employ it for lighting ordinary coaches, parlor cars, mail cars, diners, suburban trains, locomotives, railway stations, railroad yards, signal lights in towers and elsewhere, and it is also employed for trolley cars, both as to interiors and headlights. The advantage of independent headlights for trolley cars is obvious to all, when it is remembered that if the electric current chances to be cut off, as frequently occurs, away goes the headlight as well as the power, leaving car and crew in total darkness; acetylene headlights are also much cheaper than electrical ones.

For the operation of a system of this sort, a charging plant, on lines indicated in a previous chapter, is usually erected near the terminal or main railroad yard of the road to be thus equipped. The usual receiver employed on many roads for Pintsch gas contains, under ten atmospheric pressures, two hundred cubic feet of gas; it measures twenty inches in diameter and one hundred and fourteen in length. A cylinder of the same size, packed with asbestos discs and charged to forty-three per cent. with acetone, contains, at ten atmospheres, two thousand cubic feet of acetylene. This fact, together with the much greater candle power of acetylene, renders it possible to provide a car, at one charge, with gas enough to last for a month or more. Coaches have for some years been charged at St. Paul for the round trip to the Pacific coast and return, a distance of more than thirty-six hundred miles, each coach being equipped with from twelve to sixteen half-foot burners.

THE OXY-ACETYLENE BLOWPIPE.

When any gas whatever burns or undergoes combustion, in the air, it is merely combining, chemically, with the oxygen of the air. This gas, oxygen, the active portion of our atmosphere, constitutes one-fifth of the entire volume of the air, and it is only by means of it that anything burns. Whenever a candle, or an oil lamp, or a jet of illuminating gas is kindled, we see something which we are accustomed to call *flame.* This flame is nothing more or less than burning gas. Heat volatilizes, or gasifies, converts into vapor or gas, the material of the candle, or the oil in the wick of the lamp, and approaching currents of air feed the flame with oxygen by which it is sustained. Cut off the air supply and the burning ceases; increase the air supply and you may accelerate the burning. Use pure oxygen in place of air, thus getting rid of the nitrogen which constitutes the other four-fifths of the air, and you accelerate the flame still more because the oxygen is no longer diluted with four times its bulk of inert and useless material. Again, if the aim is to secure high heat, pre-heating the gases, as well as mixing them in just the right proportions at the moment of combustion, aids in accomplishing this result, and the greater the amount of fuel we can burn in a limited space and limited time, the more we can localize the heat of the combustion.

The great energy with which hydrogen combines with oxygen is turned to account for the purpose of producing high temperature by means of the instrument known as the *oxy-hydrogen blowpipe.* The separate gases are led in tubes to the burner, where they are mixed and burned to best advantage. This instrument enables us to secure a temperature of forty-five hundred degrees Fahrenheit, and it is employed in various ways. One well known application is in the familiar calcium light, or Drummond light. In this the oxy-hydrogen blowpipe is utilized to raise a piece of lime to the temperature of incandescence when it emits a powerful light, utilized for projections, for signaling at sea, etc.

Very naturally any combustible gas may replace the hydrogen and in fact coal gas is used much more commonly than hydrogen. When the attempt has been made to utilize acetylene, difficulties have been met regarding the proper adjustment of parts, which were slow to yield to inventive genius but which have been overcome in the Fouché, the Rodrigue-Ely and Gauthier torches blowpipe, are oxy-acetylene blowpipes which seem to be very perfect instruments and which are now applied to a large number of uses. They bring out the much greater calorific intensity of acetylene as compared with hydrogen and secure a temperature, locally, as high as seven thousand degrees Fahrenheit. The flame is easily regulated, all metals used in the arts are melted before it, and welding is easily accomplished. No ponderous plant is necessary; acetylene may be used right from the generator, or under pressure, as may be desired; cylinder of dissolved acetylene and of compressed oxygen are both commercial articles everywhere available, and hence a complete oxy-acetylene blowpipe outfit is readily portable.

WELDING WITH ACETYLENE.

The oxy-acetylene blowpipe affords a most valuable and practical method for welding, which far surpasses the oxy-hydrogen method and rivals electric welding. In fact, it can be employed in a large variety of work where electric welding is out of the question. It is more easily accomplished, less cumbersome and less expensive. It is stated also that it affects the character of the metal less than other forms of welding. A controlling reason why the oxy-hydrogen blowpipe cannot secure a temperature greater than four thousand to forty-five hundred degrees Fahrenheit is because oxygen and hydrogen refuse to unite at temperatures above the dissociation point of water. When acetylene is used the carbon continues to burn and produce heat, while the atmosphere of hydrogen, which is maintained about the weld, takes no part in the combustion, but prevents both the metal and the point of the blowpipe from oxidation. The work is not difficult to acquire; an intelligent workman promptly develops skill and the process is rapid. While two and a half volumes of oxygen are theoretically needed for the complete combustion of acetylene, it has been found that welding can be satisfactorily accomplished with a proportion of slightly over one and a quarter to one, hence the cost of oxygen is distinctly less than would be calculated. Since acetylene burns in oxygen with a blinding light, the operator may have occasion to protect his eyes with dark glass spectacles, but the heat of the operation, though locally intense, does not interfere with the convenience of the workman, and aside from the necessity of an occasional cooling of the torch in water, the rapidity and apparent ease of the work are not interfered with. In fact, the operation looks exceedingly simple to the casual observer, and the autogenous soldering together into homogeneous metal of considerable masses, or the cutting of thick slabs of iron goes on before his eyes with marvelous rapidity.

By reason of the facility with which it can be applied, this form of welding has become quite common abroad, and its use is being greatly extended in this country; it has already revolutionized some kinds of metal working; it is applicable in many operations which are totally impossible by the older methods; the portion of metal acted upon is so small that the factors of expansion and contraction are for the most part negligible. Welds are so homogeneous that when the piece is finished there is often no evidence that a joint has been made. Steel, iron, brass, copper and other metals used in the arts can all be treated with success, and there is authority for the statement that iron can be economically welded up to one-half inch in thickness, while any shaped hole can be cut in steel plates up to a thickness of several inches; the operator can accomplish feats impossible with the saw.

Iron or steel can be welded to much greater thickness by the process of "building up," and the writer has been reliably informed that a shaft more than twenty-four inches in diameter has been successfully welded on a steamship. For repairs in small and inaccessible rooms, on roofs and at sea, the oxy-acetylene outfit is becoming indispensable.

It will not be surprising if autogenous welding by this process largely supersedes all other methods, since it possesses such valuable advantages over all rivals in the facts that such welds are effected without any deleterious effect upon the metal, and that it is inexpensive and portable, entailing no elaborate preparations or heavy initial expenditure in regard to plant and equipment.

ACETYLENE FOR MOTIVE POWER.

An interesting field for a display of inventive genius is the application of acetylene for the generation of motive power. The high temperature of its combustion, affording a much greater explosive power than coal gas or gasoline, and the trifling trouble and expense involved in generating and storing this fuel should work together to render its use in gas engines of great benefit to the mechanical engineer. Suitably contrived engines or motors are a prime requisite and such appliances must come, but for some reason really brilliant progress along this line has been deferred. This is in spite of the fact that the efficiency of the acetylene explosion, as compared with coal gas, is about as three to one. Acetylene will eventually demonstrate its superiority in this line, perhaps is already doing so, but the difficulties seem to be slower of remedy than in the other uses to which acetylene is applied. The blowpipe was not prompt in making its appearance, but the difficulties were finally overcome, and so they will be, must be, in case of the successful acetylene motor of the future.

One interesting development in this direction is the alcohol-acetylene motor for vehicles. This invention is, in broad terms, "the process of producing a gas for power purposes which consists in carburetting air with dilute alcohol and then bringing such carburetted air into contact with calcium carbide." Such dilution of the alcohol is employed as may be determined by experiment to be most advantageous, and when this is sprayed through a layer of carbide, the water is extracted from the dilute alcohol, rendering the latter more efficient as a fuel, and converted into acetylene by reaction with the carbide. The dry gaseous mixture of alcohol vapor and acetylene passes to the explosion chamber and burns far more advantageously than alcohol vapor alone. The qualities of the two substances supplement each other, and the mixture is a fuel of considerable efficiency; it is probable that the inventive genius which is now being applied to this problem will in due time develop its full efficiency.

ACETYLENE LIGHT AND PLANT GROWTH.

The light of the acetylene flame is the nearest approach to sunlight with which we are acquainted. It is composed of the same colors and they are present in very similar degrees of intensity. Physicists have shown that if the intensity value of each color in sunlight be placed equal to unity, we have the following comparative figures:

	Sunlight.	Acetylene Light.
Red	1	1.03
Yellow	1	1.02
Green	1	0.71
Blue	1	1.46
Violet	1	1.07

The ultra-violet rays are practically absent. This composition of the light rays of acetylene naturally directed the attention of experimenters to its possible use as a plant stimulant. Careful work has been done in studying its effect on plant development at the West Virginia Experiment Station, at Cornell University and elsewhere, but chiefly at Cornell. These experiments are still under way and the results have not been fully published, but the conclusion is certain that acetylene can be used to advantage by the commercial grower in the case of certain plants. It does not possess the injurious effect of the naked electric arc and its effect is very similar to that of sunlight. It can therefore be utilized to supplement deficient sunlight, either by utilizing it at night, or as an offset to insufficient sunlight on dull and cloudy days. The forcing house equipped with acetylene is therefore able to overcome unfavorable conditions of deficient daylight. In the main, growth is increased, and the time of flower and fruit diminished; at the same time there is little or no abnormal development such as has been observed with other artificial illuminants. Acetylene can thus be used profitably to hasten a crop and prevent the delays caused by inclement weather conditions.

Some of the specific results which have been made public are worthy of mention here:

Strawberries showed an increase of twenty-five per cent. over those grown by sunlight alone, and fruit was ripened sixteen days in advance of other plants grown under like conditions of temperature but uninfluenced by acetylene.

Peas derived very marked benefit, and beans showed a noticeable gain in acetylene as compared with unaided sunlight. Tomatoes ripened more and larger fruit, though considerably later, while cucumbers exhibited no gain.

Radishes were grown in three-fourths the usual time; lettuce and other salad plants were considerably accelerated, and some of them gained in quality as well.

In general, between ninety and ninety-five per cent. of the plants which have been experimented with responded favorably to the stimulus afforded by acetylene light.

The results which have been thus far published in this field are only preliminary and there are numerous problems as yet unsolved. While many facts have been established, many tendencies shown, there is still a large and promising field for further experimentation, and it is to be hoped that the field will be further explored, for there seems to be a grand future for the profitable application of acetylene light to the wants of the market gardener and the commercial florist and horticulturist.

DRY PROCESS ACETYLENE.

The so-called Atkins dry process for generating acetylene is a method which has been exploited to some extent in England, but not employed in the United States up to the present time, so far as the writer has been able to learn. The essential feature of the system is that the gas is generated by mixing calcium carbide with soda crystals instead of with water. Soda crystals, otherwise known as sal soda or washing soda, contain a large amount of water of crystallization, about sixty-three per cent. of their total weight. The attraction of the carbide for water is so great that it extracts it from the soda crystals, and it is with this water, though originally contained in a dry substance, that the carbide reacts and produces acetylene in the usual way. Other substances which contain water of crystallization, such as alum, may be used in lieu of soda crystals, but soda is preferred. The heat accompanying the reaction is said to be very slight, not much above two hundred degrees Fahrenheit, hence distinctly below the polymerizing point of acetylene, which is above five hundred degrees Fahrenheit. The resultant gas is therefore free from impurities such as are produced by high temperatures, but it is to be doubted whether other impurities are at all affected. The residue is removed as a dry powder and contains all the actual sodium carbonate originally present in the soda crystals. The operation is not automatic and the apparatus is a mechanical one, requiring the service of an attendant and hand or other power to actuate the mixing device, which must be kept in motion whenever the gas is in the process of being generated.

It seems hardly likely that such a generator will seriously compete with the admirable automatic generators of the present day, as it involves a cost of a few cents per pound for all the water which takes part in the decomposition of the carbide, with no offsetting advantage in the salable character of the residuum, increased value of the resultant gas, or otherwise.

CALCIUM CYANAMID.

One of the most important facts about calcium carbide is that discovered by two German chemists, Frank and Caro, namely, its property of absorbing nitrogen at a high temperature. Because of the great need of soluble nitrogen compounds for application to the soil as fertilizers for the growing of plants of every sort, every known source of such nitrogen compounds has been drawn upon to the fullest extent during recent years. All the agriculturally available nitrogen has hitherto been drawn from only three chief sources: ammonium salts resulting from the by-products of the illuminating gas industry, chiefly sulphate and muriate; Chilean saltpeter, a substance secured from the rainless district of western South America, and nitrogenous residues from the slaughter house, such as dried blood and tankage. These three sources are drawn upon to their utmost capacity and the price of the nitrogen-containing fertilizers remains so high that the discovery of a new source of nitrogen available for the purpose is a thing greatly to be desired. The atmosphere of the earth supplies us with never failing quantities of the desired element nitrogen, containing as it does nearly eighty per cent., whether by weight or volume. Unfortunately, however, this great superabundance of nitrogen, within reach of everybody and at no cost, is free or uncombined, and hence incapable of being utilized for the purpose desired. Free nitrogen will not answer the requirements of a fertilizer; nature requires that it be in the form of soluble compounds in order to be available. It is not to be wondered at, therefore, that it has been the aim of much expensive and laborious investigation, during many years, to find a practical means of compelling nitrogen of the air to enter into unwilling union with some other elements so as to yield soluble nitrogenous compounds in paying quantities. Every field in which there is any promise of ultimate success has been thoroughly explored. The best scientific ability in the world has been lavished upon this problem and no expense has been spared. One process after another has been proposed, tested out, and abandoned as unsuccessful. Finally, within the last six or seven years, the discovery above referred to has been the subject of earnest, active collaboration on the part of engineers, agriculturists and chemists, with the result that factories are now established and engaged in the production of a new nitrogenous fertilizer made by the fixation of nitrogen along entirely new lines, by the aid of calcium carbide. The fundamental reaction is as follows:

CaC_2	$+$	N_2	$=$	$CaCN_2$	$+$	C
Calcium carbide	and	nitrogen	yield	calcium cyanamid	and	carbon,

which means that calcium carbide, heated with nitrogen, under suitable conditions, yields about its own weight of calcium cyanamid, frequently called lime-nitrogen, and a certain amount of carbon. The method of bringing about the reaction is not in the least complicated. The carbide is placed in electrical furnaces of especial form, and nitrogen brought into contact by suitable pipes. The nitrogen may be secured from the atmosphere in any known way, and for this purpose have served both the old copper method, whereby metallic copper is oxidized in air until the oxygen is all removed, leaving only the nitrogen, and the more recent Linde process, in which air is first liquefied and then separated by fractional distillation into its two constituents.

The lime-nitrogen is removed from the furnace, reduced to a fine powder, delivered from the factory in ordinary jute bags, and is immediately ready for use. It possesses a dark gray color and a slight odor resembling sulphuretted hydrogen; it is absolutely stable in its constitution, non-hygroscopic and unaffected by exposure to air or moisture. It is sufficiently soluble to answer to the requirements in that respect, and it is absolutely fireproof, hence produces no insurance hazard.

The question of its fitness for agricultural purposes has been thoroughly investigated by European agriculturists and chemists, and all results are favorable to its employment. It has now passed beyond the experimental stage and is widely employed in half a dozen different countries, so that its adaptability to many soils, different climates, and a wide variety of crops is assured. There can be no further question as to its fitness for use as a fertilizer, and it ranks in value fully up with Chile saltpeter and ammonium sulphate, in some individual cases surpassing these.

It is possible to make ammonia from cyanamid, and as there is a large demand for this commodity, both as liquid ammonia for refrigerating machinery and for the production of ammoniacal salts, and as its production from calcium cyanamid seems to be perfected on an industrial scale, a large use of carbide may here be indicated.

It is possible also to convert cyanamid into the various metallic cyanides, and as there is a considerable market for these, particularly for cyanide of potassium, which finds wide application in the recovery of precious metals from low grade ores and tailings, there may be in this direction also a coming market for carbide.

Cyanamid may also be utilized in the preparation of other products in considerable number, rather too technical in character for enumeration in this paper except one, namely, nitric acid. The nitrogen contained in calcium cyanamid can unquestionably be transformed into nitric acid and commercial nitrates; investigators are engaged in solving this problem at the present time, but we must patiently await the careful elaboration of this very promising process before we can expect to see it placed on a commercial basis.

THE UNIVERSALITY OF ACETYLENE.

Prometheus, in mythological lore, stole fire from heaven and brought it down to earth for the use of man. The beast of the field had been man's enemy; Prometheus' torch, celestial gift, lighted at the chariot of the sun, enabled man to forge weapons with which to overpower him. Man had had no heat; with heaven-stolen fire he warmed his dwellings and thenceforth inhabited every clime. With it also he coaxed the metals from their hiding places beneath the earth, established the arts and manufactures, coined money and opened the avenues of commerce. Jupiter, 'tis said, from his throne on old Olympus, beheld this fire upon earth, inquired what it meant, and when informed, his rage knew no bounds! But fire burns on, from day to day, from age to age, and "tries every man's work of what sort it is."

Once more, this time within a dozen years, has the chariot of the sun again, even more truthfully than before, yielded a bit of itself to our nineteenth century Prometheus, fitting climax to a century prolific in marvels! No need of mythology now, nor need Jupiter rage again! Acetylene has penetrated to every civilized land. It lights the path of man where'er he goes. Flowers turn toward it as toward the sun itself, and man knows no other illuminant so like the heavenly luminary.

The completed list of the possible useful applications of calcium carbide and acetylene has not been made. So fruitful are these marvelous substances in the development of new compounds like cyanamid, not heretofore known except as a curiosity, of new industries like the industry of dissolved acetylene, purely an invention of the last decade, that we no longer look on with wonder when something new is proposed.

We have but to glance at one of the journals devoted to the acetylene industry, and we read, on one page, of the manufacture of *tetra-chlor-acetylene,* a derivative of acetylene made by the direct action of chlorine upon it, and advertised as a cheap and excellent solvent for organic substances, chlorine, phosphorus, fats and resins. The substance is described as an oil, possessing an aromatic odor, boiling at two hundred and ninety-six degrees Fahrenheit, and if it can be sold at a price which places it in competition with other fat solvents, an additional use for acetylene will here be found.

On another page we read of several patents which have been taken out by a prominent German chemist for processes of manufacturing pure hydrogen by the aid of calcium carbide. Still further on we are instructed as to the preparation of *argon,* by the assistance of calcium carbide. Elsewhere we find a description of an oxy-acetylene crucible furnace for facilitating chemical operations, and again of the synthesis of alcohol and related substances from acetylene, and still again of the manufacture of lamp-black from the smoke of the acetylene flame.

Indeed, the chemist finds acetylene an indispensable adjunct to other reagents in a well-equipped chemical laboratory; he requires it for its perfect light for polariscopic work and for reading many end-reactions in volumetric analysis, as well as for developing the highest heat available in the ordinary chemical laboratory, and for the occasional precipitation of certain metals. Again the chemist finds in calcium carbide a most efficient desiccating agent, its marked avidity for water rendering it impossible for any moisture to exist in contact with it.

Nor are all these developing uses of acetylene and carbide merely of a chemical character. We find acetylene growing in popularity for use with the heliograph for signaling on land and sea, for surgical uses in army and navy, where its penetrating character and correct exposition of colors render it invaluable. We know not only that carbide is one of the most useful agents for combating the deadly phylloxera and other parasites of the vine, but that acetylene itself is employed as an insecticide with no inconsiderable success.

The heat of the oxy-acetylene blowpipe flame has been utilized for the preparation of artificial gems, though probably, as yet, with a measure of success less brilliant than the gems themselves. In a more humble way, acetylene heat is employed in the branding of meats, and acetylene light for testing eggs.

To-night a million acetylene lights are burning in American homes; thousands of acetylene searchlights scan the water from yachts and sailing craft of every description; thousands of automobiles are guided by the aid of penetrating acetylene lights; hundreds of vessels steer by the light of acetylene beacons. Passengers on railroad trains are enjoying their evening meal and reading the evening papers under the soft light of our beautiful illuminant, while locomotive engineers drive their trains into veritable tunnels, hollowed out of darkness and fog, by means of brilliant and efficient acetylene headlights, the equal of which has never before been known in the history of railroading.

Five hundred towns and villages derive their public and private illumination from central plants where acetylene is generated; two hundred thousand horsepower, developed by harnessing waterfall and mountain torrent, are constantly absorbed in many factories throughout the world for the production of carbide, while as many tons of the marvelous product are yearly poured out from fiery furnaces which grumble and roar under the fierce strain.

Millions of capital are profitably invested in no less than a dozen different countries in the production of carbide and acetylene and in its employment for light, heat and power; in autogenous welding by the aid of the oxy-acetylene blowpipe; in dissolved acetylene; in calcium cyanamid; and in all the lesser ramifications of the general industry which it has been the province of this pamphlet to point out.

In short, so many, and so varied, and so widely applicable are the useful applications of acetylene and calcium carbide in aiding the operations of man, that we may truthfully say that acetylene has most decidedly contributed a large share, as did Prometheus's fire of old, to man's dominion over the earth.

BIBLIOGRAPHY.

Following is a list of the principal published works on calcium carbide and acetylene:

Das Acetylen. J. H. Vogel, Berlin, 1900.

Acetylene. F. H. Leeds and W. J. A. Butterfield. Philadelphia, 1903.

Ueber Acetylenbeleuchtungsanlagen. J. Schultze. Leipsic, 1901.

Die Acetylen- und Calciumcarbidindustrie vom patentlichen Standpunkt. J. Ephraim. Halle, 1900.

Acetylene Gas, How to Make and Use It. C. N. Turner. New York, 1903.

Acetylene, a Handbook for the Student and Manufacturer. V. B. Lewes. London, 1900.

Acetylenkalender für 1900. Leipsic, 1899.

Das Acetylen in der Technik. F. B. Ahrens. Stuttgart, 1899.

Acetylenzentralen. J. H. Vogel. Halle, 1901.

Anleitung zur sicherheitstechnischen Prüfung und Begutachtung von Acetylen. N. Caro. Berlin, 1902.

Calciumcarbid und Acetylen. Viebeck. Halle, 1897.

Calciumcarbid und Acetylen. Liebetanz. Leipsic, 1898.

Ueber Calciumcarbid und Acetylengas. Erdmann. Leipsic, 1895.

Die Calciumcarbid Industrie. Wangemann. Dresden, 1904.

Calciumkarbid und Acetylen. F. Dommer. Munich and Leipsic, 1898.

Calciumcarbid und Acetylen in Vergangenheit, Gegenwart und Zukunft. Leipsic, 1897.

Le Carbure de Calcium et l'Acétylène. C. de Perrodil. Paris, 1897.

l'Éclairage à l'Acétylène. Capelle. Paris, 1897.

Elektrochemische Untersuchungen am Acetylen. Vienna, 1902.

Führer durch die gesammte Calciumcarbid- und Acetylen-Literatur. A. Ludwig. Berlin, 1899.

Handbuch für Acetylen in technischer und wissenschaftlicher Hinsicht. Caro, Ludwig und Vogel. Braunschweig, 1904.

Hauptmomente der Acetylen und Carbid Industrie. E. Fernderl. Vienna, 1900.

Die Herstellung, Aufbewahrung und Verwendung von Acetylengas und Lagerung von Carbid. J. Knappich. Halle, 1902.

Hilgsbuch für Installationen von Acetylenbeleuchtungsanlagen. Leipsic, 1900.

Jahrbuch für Acetylen und Carbid. Altschul, Scheel und Vogel. Halle, 1900.

Kalender für Acetylentechniker. Schaefer. Berlin, 1899.

Zur Kentniss des Acetylens, etc. Heidelberg, 1892.

Katalog der zweiten Acetylen Fachausstellung. Cassel, 1899.

Lighting by Acetylene. Frederick Dye. New York, 1902.

Lighting by Acetylene. W. E. Gibbs, New York, 1898.

Methoden zur Bestimmung der Gasausbeute aus Calciumcarbid. Herausgegeben vom deutschen Acetylenverein. Halle, 1900.

Die neue Concurrenten des Steinkohlengases. (Acetylen, Wasergas, Luftgas.) Schaefer. Munich, 1900.

Praktisches Handbuch der Acetylenbeleuchtung und Calciumcarbidfabrikation. A. Ludwig. Berlin, 1898.

Some Reactions of Acetylene. J. A. Nieuwland. Notre Dame, Ind., 1904.

Stand und Zukunft der Acetylenbeleuchtung. Froehlich und Herzfeld. Berlin, 1895.

Taschenbuch für Acetylentechniker. Ludewig. Berlin, 1900.

Vormen des deutschen Acetylenvereines. Halle, 1901.

Vorschriften für Herstellung und Benutzung von Acetylen. Liebetanz. Leipsic, 1899.

INDEX TO PERIODICAL LITERATURE.

List of Journals Reviewed.

American Architect	Weekly, New York	Am. Arch.
American Engineer and Railroad Journal	Monthly, New York	Am. Eng. & R. R. Jour.
American Gas Lighting Journal	Weekly, New York	Am. Gas Ltg. Jour.
American Institute of Electrical Engineers	Monthly, New York	Am. Inst. Elec. Eng.
American Journal of Science	Monthly, New Haven	Am. Jour. Sci.
American Manufactures and Iron World	Weekly, Pittsburg	Am. Mfr. Ir. Wld.
Annales des Ponts et Chaussés	Weekly, Paris	Ann. d. Ponts et Chaussés
Architect	Weekly, London	Arch't.
Automobile	Monthly, New York	Auto.
Builder	Weekly, London	Bldr.
Canadian Electrical News	Monthly, Toronto	Can. Elec. News.
Canadian Engineer	Monthly, Toronto and Montreal	Can. Eng.
Cassier's Monthly	Monthly, New York and London	Cassier's.
Colliery Guardian	Weekly, London	Col. Guard.
Compt. Rend. de l'Académie des Sciences	Weekly, Paris	Comptes Rendus.
Consular Reports	Monthly, Washington	Cons. Rep.
Cosmopolitan Magazine	Monthly, New York	Cosmop.
Country Life	Monthly, New York	Count. Life.
Domestic Engineering	Weekly, Chicago	Dom. Eng.
Electrical Engineer	Weekly, London	Elec. Eng.
Electrical Review	Weekly, New York	Elec. Rev.
Electrical Review	Monthly, London	Elec. Rev., Lond.
Electrical World	Weekly, New York	Elec. Wld.
Electrician	Weekly, London	Elect'n.

Electricity	Weekly, New York	Elec.
Electrochemical and Metallurgical Industry	Monthly, New York	Electrochem. Ind.
Electro-Techniker	Bi-monthly, Vienna	Elec. Tech.
Electrochemische Zeitschrift	Monthly, Berlin	El'ch. Zeit.
Elettricita	Weekly, Milan	Elet.
Engineer	Weekly, London	Engr.
Engineering	Weekly, London	Engng.
Engineering Magazine	Monthly, New York and London	Eng. Mag.
Engineering News	Weekly, New York	Eng. News.
Engineering Record	Weekly, New York	Eng. Rec.
Engineers' Society of Western Pennsylvania	Monthly, Pittsburg	Eng. Soc. W. Pa.
Fire and Water	Weekly, New York	
Gas World	Weekly, London	Gas Wld.
Génie Civil	Weekly, Paris	Gen. Civ.
Génie Moderne	Weekly, Paris	Gen. Mod.
Gesundheits Ingenieur	Semi-monthly, Munich	Gesund. Ing.
Glaser's Annalen für Gewerbe und Bauwesen	Semi-monthly, Berlin	Glas. Ann.
Glückauf	Weekly, Essen, Germany	Glück.
Harper's Weekly	Weekly, New York	Harp. Wkly.
Heating and Ventilating	Monthly, New York	Heat. & Vent.
Horseless Age	Weekly, New York	H. Age.
Illustrated Carpenter and Builder	Weekly, London	Ill. Car. & Build.
Illustrirte Zeitscrhift. für klein und Strassenbahnen	Semi-monthly, Berlin	Ill. Zeit. Strass.
Indian and Eastern Engineer	Monthly, Calcutta	Ind. & East. Eng.
Industries and Iron	Weekly, London	Ind. & Ir.
De Ingenieur	Weekly, Hague	Ing.
Iron Age	Weekly, New York	Ir. Age.
Iron and Coal Trades Review	Weekly, London	Ir. Coal. Tr. Rev.
Journal of the Association of Engineering Societies	Monthly, Boston	Jour. Assn. Eng. Soc.
Journal Gas Lighting	Weekly, London	Jour. Gas Ltg.
Journal Franklin Institute	Monthly, Philadelphia	Jour. Fr. Inst.
Journal of the Society of Arts	Weekly, London	Jour. Soc. Arts.
Journal of the Society of Chemical Industry	Semi-monthly, London	Jour. Soc. Chem. Ind.
Knowledge	Weekly, London	Knowl.
Ladies' Home Journal	Monthly, Philadelphia	Lad. Home Jour.
Mechanical Engineer	Weekly, Manchester, Eng.	Mech. Eng.
Mémoirs de la Société des Ingenieurs Civils de France	Monthly, Paris	Mem. Soc. Ing. Civ.
Metal Worker	Weekly, New York	Met. Work.
Mining and Scientific Press	Weekly, San Francisco	Min. & Sci. Press.
Municipal Journal and Engineer	Monthly, New York	Munic. Jour. & Eng.
Nature	Weekly, London	
New Science Review	Quarterly, New York	New Sci. Rev.
Oesterreichische Monatschrift für den Oeffentlichen Baudienst	Monthly, Vienna	Oest. Monat.
Oesterreichische Wochenschrift für den Oeffentlichen Baudienst	Weekly, Vienna	Oest. Woch.
Oesterreichische Zeitschrift für Berg und Hüttenwesen	Weekly, Vienna	Oest. Zeit. Berg. Hütt
Plumber and Decorator	Monthly, London	Plum. & Dec.
Popular Science	Monthly, New York	Pop. Sci.
Practical Engineer	Weekly, London	Prac. Eng.
Proceedings Iowa Railway Club		Pro. Iowa Ry. Club.
Progressive Age	Semi-monthly, New York	Pro. Age.
Railroad Gazette	Weekly, New York	R. R. Gaz.
Railway Age	Weekly, Chicago	Ry. Age.
Railway and Engineering Review	Weekly, Chicago	Ry. & Eng. Rev.
Revue Général des Sciences	Weekly, Paris	Rev. Sci.
Revue Technique	Semi-monthly, Paris	Rev. Tech.
Revue Universelle des Mines	Monthly, Liége	Rev. Univ. d. Mines.
Sanitary Plumbing	Semi-weekly, New York	San. Plumb.
Schweizerische Bauzeitung	Weekly, Zurich	Schw. Bauz.
Science	Weekly, New York	Sci.
Scientific American	Weekly, New York	Sci. Am.
Scientific American Supplement	Weekly, New York	Sci. Am. Sup.
Sibley Journal of Engineering	Monthly, Ithaca, N. Y.	Sib. Jour.
Stahl und Eisen	Semi-weekly, Düsseldorf	S. und E.
Stevens Institute Indicator	Quarterly, Hoboken, N. J.	Stev. Inst. Ind.
Technology Quarterly	Mass. Institute of Technology, Boston	Tech. Quar.
Western Electrician	Weekly, Chicago	West. Elec.

Zeitschrift für Elektrochemie.	Weekly, Halle	Zeit. El'ch.
Zeitschrift des Oesterreichischen Ingenieur und Architekten Vereines	Weekly, Vienna	Zeit. Oest. Ing.
Zeitschrift des Vereines Deutscher Ingenieure......	Weekly, Berlin	Zeit. Deutsch. Ing.

Index.

Die Acetylen-Centrale für die Beleuchtung von Schönsee. Paul Wolff. Glas. Ann., June 1, 1899.
Acetylene Coach Lighting. Ry. & Eng. Rev., Apr. 21, 1906.
Acetylene v. Coal Gas or Electricity. H. Allen. Sci. Am. Sup., Dec. 1, 1906.
Acetylene Congress at Frankfort. Jour. Gas Ltg., Nov. 15, 1898.
Acetylene Congress at Nuremberg. Engr., Lond., Nov. 17, 1890.
l'Acétylène, Au Concours Agricole. Gén. Mod., May 1, 1897.
l'Acétylène Dissous. Rev. Tech., Mar. 25, 1897.
l'Acétylène Dissous et ses Principales Applications. A. Janet, Gén. Civ., July 18, 1903.
l'Acétylène Dissous et ses Principales Applications. F. Bourdil. Mem. Soc. Ing. Civ., Nov., 1903.
Acetylen, Einige Weitere Mittheilungen ueber. Glas. Ann., Mar. 15, 1897.
Acetylene for Enriching Gas Produced from Lean Coals. T. Vautier. Pro. Age, Dec. 1, 1897.
Acetylene versus Ethylene. Engr., Dec. 16, 1898.
Acetylene Exhibition and Congress at Budapest. Engr., Aug. 11, 1898.
Acetylene Explosions. Editorial. Am. Gas Ltg. Jour., Feb. 14,
Acetylene Explosions. Sci. Am., Feb. 5, 1898.
Acetylene Explosions. Editorial. Am. Gas Lgt. Jour., Feb. 14, 1898.
Das Acetylen und seine Explosions-Gefährlichkeit. Glas. Ann., Apr. 1, 1897.
Acetylene as an Explosive. Engr., Oct. 23, 1896.
Acetylene Flame. E. L. Nichols. Jour. Fr. Inst., Nov., 1900.
Acetylene Gas. W. R. Addicks. Am. Gas Ltg. Jour. Mar. 9, 1896.
Acetylene Gas. George Black. Can. Elec. News, July, 1896.
Acetylene Gas. T. Glover. Gas Wld., Feb. 26, 1898.
Acetylene Gas. V. B. Lewes. Ill. Car. & Build., Feb. 10, 1899.
Acetylene Gas. D. P. Heap. Cosmop., Sept., 1901.
Acetylene Gas. H. C. Earle. Sib. Jour., Nov., 1901.
Acetylene Gas. Gas Wld., Feb. 1, 1902.
Acetylene Gas Accidents. Elec. Rev., Apr. 28, 1899.
Acetylene Gas and Calcium Carbide. G. F. Thompson, London, 1898.
Acetylene Gas on a Canadian Railway. Can. Eng., Apr., 1898.
Acetylene Gas in the Chassiron Light House. Engng., Oct. 16, 1903.
Acetylene Gas Considered Commercially. H. Allen. Elec. Rev., Jan. 5, 1900.

Acetylene Gas Considered from the Fireman's Standpoint. Wm. McDavitt. San. Plumb., Oct. 2, 1899.
Acetylene Gas, Cost of. Editorial. Pro. Age, Apr. 1, 1899.
Acetylene for Gas Engines. Engr., Jan. 17, 1902.
Acetylene Gas Exhibition. London. P. F. Nursey, Ind. & Ir., June 17, 1898.
Acetylene Gas Exhibition. London. Col. Guard., July 1, 1898.
Acetylene Gas Generator. J. H. Adams. Sci. Am., Apr. 23, 1898.
Acetylene Gas Generator, A New. Eng. News, Oct. 4, 1900.
Acetylene Gas Generator for Locomotive Lighting. R. R. Gaz., Oct. 11, 1901.
Acetylene Gas in Germany and Sweden. Cons. Rep., Oct. 31, 1901.
Acetylene Gas Hazard. L. J. Kram. Sci. Am., Oct. 8, 1898.
Acetylene Gas Lamp for Bicycles. Sci. Am., Nov. 13, 1897.
Acetylene Gas for Lighting. A. Lipschutz. Am. Arch., Aug. 11, 1900.
Acetylene Gas for Lighting. H. L. Taylor. Dom. Eng., Sept., 1899.
Acetylene Gas for Locomotive Headlights. Ry. & Eng. Rev., Mar. 4, 1899.
Acetylene Gas, The New Illuminant. C. T. Marshall. Arch't., Mar. 11, 1898.
Acetylene Gas Plant of the Great Northern at St. Paul. R. R. Gaz., June 12, 1903.
Acetylene Gas Plant at Schönsee, Prussia. Eng. Rec., Oct. 21, 1899.
Acetylene Gas, Report Upon. Zeit. Oest. Ing., Feb. 26, 1897.
Acetylene and Gasoline. L. Berger. H. Age, June 20, 1900.
Acetylene Generators. Abstracts. V. B. Lewes. Jour. Gas Ltg., June 14, 1898.
Acetylene Generators. N. Caro. Abstract trans. Jour. Gas Ltg., June 14, 1898.
Acetylene Generator, A New. Sci. Am. Sup., Aug. 26, 1899.
Acetylene Generators for Motor Vehicles. W. F. Cooper. H. Age, June 20, 1900.
Acetylene in German Villages. Engr., May 13, 1904.
Acetylene in Germany. A. Rose. Engr., Dec. 14, 1900.
Acetylen Gruben-Sicherheitslampe, System Stuchlik. Oest. Zeit. Berg Hütt., Jan. 10, 1903.
Acetylene for Headlights. Ry. & Eng. Rev., Mar. 4, 1899.
Acetylene Headlights on Locomotives. G. L. Wilson. Ry. Age, Oct. 11, 1901.
Acetylen von Hygienischen Standpunkte. Joseph Verters. Gesund. Ing., July 31, 1898.

Acetylene Illumination. Pro. Age, May 15, 1907.
The Acetylene Industry. G. T. Holloway. Knowl., Apr. & May, 1899.
Acetylene Installation at Wabash, Ind. Am. Gas Ltg. Jour., Nov. 28, 1898.
Acetylene Laboratory Burner. A. Lachman. Pro. Age, Aug. 15, 1900.
Acetylene Lamps for Miners' Use. K. Baumgartner. Col. Guard., Feb. 23, 1900.
Acetylene for Lighting. A. Lipschutz. Am. Arch., Aug. 11, 1900.
Acetylene Lighting. F. Cortez Wilson. San. Plumb., Dec. 1, 1899.
Acetylene Lighting at Hawes, Eng. Engr., Oct. 13, 1899.
Acetylene Lighting Plants for Villages. R. C. Miller. Met. Work., May 4, 1901.
Acetylene Lighting and Storage. R. E. Bruckner. Stev. Inst. Ind., Oct., 1907.
Acetylene Lighting at Vienna. Pro. Age, Mar. 15, 1898.
Acetylene and the Manufacture of Mineral Blacks. F. Jaubert. Sci. Am. Sup., Mar. 17, 1900.
Acetylene Manufacture, Some Theoretical Aspects of. Engr., June 24, 1898.
Acetylene as a Mine Illuminant. A. Holland. Ir. Coal Tr. Rev., Apr. 28, 1899.
Acetylen und Mischgasanstalten für Beleuchtung der Eisenbahnwagen. Glas. Ann., June 15, 1898.
Acetylene and Its Modern Developments. V. B. Lewes. Pro. Age, Aug. 15 and Sept. 1, 1898.
Acetylene for Motors. Abs. trans. from A. v. Ihering. Jour. Gas Ltg., Dec. 15, 1896.
Acetylene, the New Illuminant. V. J. Youmans. Pop. Sci., Oct., 1896.
Acetylene Plant for House Supplies. Jour. Gas Ltg., Oct. 4, 1898.
Acetylene for the Production of Motive Force. P. Chevillard. Pro. Age, Oct. 15, 1897.
Acetylene for Railroad Lighting. R. R. Gaz., Sept. 28, 1900.
Acetylene for Railway Lighting in France. Engr., Mar. 4, 1902.
Acetylene Railway Station and Train Lighting. A. Lipschutz. Jour. Assn. Eng. Soc., June, 1900.
Acetylene Regulations and Insurance. Met. Work., Dec. 11, 1898.
Acetylene Regulations for New York. Pro. Age, Mar. 1, 1898.
Acetylene Safer Than Most Other Illuminants. H. F. Fuller. Dom. Eng., June, 1899.
Acetylene Search Lights for Automobiles. Sci. Am., Nov. 15, 1902.

The Acetylene Situation. Editorial. Pro. Age, Sept. 1, 1899.

Acetylen u. Spiritus Glühlichtbeleuchtung. Glas. Ann., Feb. 15, 1898.

Die Acetylen Stadtanlagen in Oliva, Prussia. Paul Wolff. Ill. Zeit. Strass., Jan. 16, 1899.

The Acetylene Standard for Light. H. E. Clifford & J. S. Smyser. Tech. Quar., Dec., 1898.

Acetylene Standard for Photometry. Trans. from M. Voille. Pro. Age, Mar. 16, 1896.

Acetylene, Stored and Transported in Safety. J. S. Seymour. Jour. Fr. Inst., July, 1903.

Acetylene Stoves and Their Application. A. J. Murdock. Met. Work., July 6, 1901.

Acetylene, Success of. Country Calendar. June, 1905.

Acetylene for Train Lighting. A. Lipschutz. Am. Gas Ltg. Jour., Sept. 17, 1900.

Acetylene from the Underwriters' Standpoint. Can. Engr., Jan., 1898.

Acetylene, Use of. E. Renouf. Pop. Sci., July, 1899.

Acetylene, the Use of. A. H. Milliken. West. Elec., Oct. 9, 1897.

Acetylene Used in Raising Sunken Vessels. Sci. Am., Apr. 28, 1900.

Action of Acetylene on the Oxides of Copper. F. A. Gooch & D. F. Baldwin. Am. Jour. Sci., Nov., 1901.

The Adlake Acetylene Car Lighting System. L. K. Gilson. Ry. & Eng. Rev., Feb. 28, 1898.

The Adlake System of Acetylene Gas for Car Lighting. R. R. Gaz., Nov. 22, 1901.

Advantages of Acetylene. Can. Eng., Apr., 1903.

Alcohol Acetylene Mixture for Internal Combustion Engines. Sci. Am., Mar. 23, 1907.

Analyse de quelques Echantillons Industriels de Carbure de Calcium. Henri Moissan. Compt. Rend., Oct. 3, 1898.

Analysis of Calcium Carbide and Acetylene and Purification of Latter. Gas Wld., Jan. 1, 1898.

Analysis and Purification of Acetylene. Rossel and Landriset. Jour. Gas Ltg., June 11, 1901.

Annuaire Internationale de l'Acétylène. Paris. A periodical, annual since 1907.

Die Anwendung der Acetylen-beleuchtung für Eisenbahnzwecke. Paul Wolff. Ill. Zeit. Strass., Oct. 16, 1899.

Apparatus for Lighting with Acetylene. Camp and Mareschal. Sci. Am. Sup., Feb. 27, 1897.

Apparatus for the Production of Acetylene Gas. Sci. Am. Sup., June 18, 1898.
Apparatus for the Production of Acetylene Gas. Sci. Am. Sup., July 2, 1898.
Appareils Cousin à Chute Automatique de Carbure. Rev. Tech., Oct. 25, 1898.
The Application of Acetylene Illumination to Country Homes. G. G. Pond. Bulletin No. 57. Published by the Pa. Department of Agriculture, Harrisburg, 1899.
Application of Acetylene to Lighting. L. M. Bullier. Pro. Age, Nov. 1, 1897.
Sur l'Aptitude de l'Acétylène Melangé à des Gas Inertes. Berthelot and Vieille. Compt. Rend., Mar. 27, 1899.
Artificial Illumination for Ordinary Business and Domestic Purposes. Am. Gas Ltg. Jour., Aug. 7, 1899.
Artificial Light. Modern Methods Compared. D. S. Jacobus. Pro. Age, May 15, 1897.
Atkins Dry Process of Generating Acetylene Gas. Sci. Am. Sup., Oct. 20, 1906.
Autogene Schweissung. S. und E., Aug. 1, 1905.
Autogenous Lead Soldering. M. U. Schoop. Electrochem. Ind., July, 1905.
Autogenous Welding. Augustine Davis. Met. Work., July 20, 1907.
Autogenous Welding of Metals by the Oxy-Acetylene Blowpipe and a New Method of Generating Oxygen. Andre Beltzer. Electrochem. Ind., July, 1906.
Autogenous Welding with Oxy-Acetylene High Pressure Blowpipe. Sci. Am., Mar. 23, 1907.
An Automatic Generator. Sci. Am., Sept. 15, 1900.
Baldwin Acetylene Lamp. Sci. Am., June 23, 1900.
Bericht über den corporativen Besuch der von der Acetylengas Gesellschaft Veranstalteten Demonstration. Zeit. Oest. Ing., Feb. 26, 1897.
Die Benutzung der Hochofengase zur Fabrikation von Calcium Carbid. Glück., Nov. 10, 1900.
Berlin Acetylene Exposition. Pro. Age, Apr. 15, 1898.
Best Methods of Generating Acetylene. J. A. Painchaud. Ir. Age, Oct. 25, 1900.
British Carbide Acetylene Patents. Interview with V. B. Lewes. Gas Wld., Apr. 11, 1896.
Budapest Acetylene Exposition. Sci. Am. Sup., Apr., 1900.
Burners for Acetylene. Abstract translation from Paul Wolff. Jour. Gas Ltg., Mar. 1, 1898.

Calcic Carbide Plant at Niagara Falls, O. E. Dunlap. West. Elec., Jan. 18, 1896.

Calcium Carbide and Acetylene. Am. Gas Ltg. Jour., page 903, 1894.

Calcium Carbide and Acetylene. W. H. Wahl. Jour. Fr. Inst., Jan., 1895.

Calcium Carbide and Acetylene. Can. Eng., Jan., 1896.

Calcium Carbide and Acetylene. Eng. News, Jan. 30, 1896.

Calcium Carbide-Acetylene Gas. Pro. Age, Apr. 15, 1896.

Calcium Carbide and Acetylene. C. H. Day. Cons. Rep., May 24, 1897.

Calcium Carbide and Acetylene. Editorial. Am. Gas Ltg. Jour., Apr. 4, 1898.

Calcium Carbide and Acetylene. Serial, Ind. & Ir., Apr. 14, 1898.

Calcium Carbide and Acetylene. Min. Ind., Vol. VI, page 65, 1898.

Calcium Carbide and Acetylene. Electrochem. Ind., Dec., 1902.

Calcium Carbide and Acetylene and Their Applications. E. Hospitalier, Pro. Age, May 1, 1896.

Calcium Carbide and Acetylene in France. Thackara. Cons. Rep., Aug. 4, 1898.

The Calcium Carbide and Acetylene Gas Industry. J. B. C. Kershaw. Elec. Rev., July 7, 1899.

Calcium Carbide and Its Commercial Development. Sci. Am., June 11, 1904.

Calcium Carbide at the Congress of Applied Chemistry. Paris. Sci. Am., Oct. 20, 1900.

Die Calcium Carbid Fabrication und deren Zuzammenhang mit der Eisenindustrie. H. Liebetanz. Serial. S. und E., Mar. 1, 1900.

Calcium Carbide Industry. Engng., July 15 and Dec. 6, 1898.

Das Calcium Carbid als Mittel zur Arbeitsübertragung. E. Neuberg. Ee'ch. Zeit., Mar., 1900.

Calcium Carbide at Neuhausen and Bitterfield. Pro. Age, Jan. 1, 1898.

Calcium Carbide in Ontario. From the Report of A. Blue. Ir. Age, Apr. 7, 1898.

Calcium Carbide Process. U. S. Patent No. 541137. June 18, 1895.

Calcium Carbide Works at Ingleton. F. J. A. Matthews. Elect'n, Sept. 9, 1898.

Calcium Cyanamid, Commercial and Physical Qualities. A Report by Chas. E. Munroe, Washington, June 1, 1907.

Calcium Cyanamid (Lime Nitrogen) from the Standpoint of American Fertilizer Manufacturer. A Report by C. Glaser, Baltimore, June 1, 1907.

Carbide of Calcium and Acetylene. Engr., June 15, 1897.

Carbide of Calcium and Acetylene. Translation of C. de Perrodil's Book "Le Carbure de Calcium et l'Acétylène." Serial. Pro. Age, Apr. 1, 1898.

Carbides and Acetylene, Commercially Considered. T. L. Willson and J. J. Suckert. Jour. Fr. Inst., Apr., 1895.

Carbide Plant at Niagara Falls. Am. Gas Ltg. Jour., May 8, 1899.

Das Carbidwerk Flums. Schw. Bauz., Sept. 14 and 28, and Oct. 5 and 12, 1901.

Carbolite. Editorial Pro. Age, Aug. 15, 1898.

Car Lighting by Acetylene Gas on the Avery-Acetylite System. Eng. News, Dec. 5, 1901.

Car Lighting. S. T. Canfield. R. R. Gaz., July 11, 1902.

Car Lighting by Acetylene and Electricity. S. T. Canfield. Am. Engr. & R. R. Jour., Aug., 1902.

Car Lighting. E. G. Fisher. Ry. & Eng. Rev., Feb. 14, 1903.

Care of Lamps and Generators. Wm. Gray. Harp. Wkly., Mar. 16, 1907.

Le Chalumeau Oxy-Acétylènique. E. Fouché. Gén. Civ., Sept. 26, 1903.

The Chloride of Lime Process for the Purification of Acetylene. Abst. trans. from F. B. Ahrens. Jour. Gas Ltg., Sept. 19, 1899.

Commercial Value of Acetylene Gas as an Illuminant. T. A. Ferguson. Eng. News, May 14, 1896.

Comparative Economy of Various Methods of Illumination. Sci. Am. Sup., Jan. 12, 1907.

Compte Rendu Officiel de la Deuxième Session du Comité International du Carbure et de l'Acétylène, Tenue à Paris, December, 1906.

Compt. Rend. Officiel de la Première Session du Comité International du Carbure de l'Acétylène Tenue à Nürnberg, June, 1906.

Concours d'Appareils à Acétylène. Paris, 1907.

Conditions Existing in Generators. V. B. Lewes. Pro. Age, Aug. 15 and Sept. 1, 1898.

Conditions of Temperature Existing in Acetylene Generators. V. B. Lewes. Jour. Gas Ltg., March 17, 1899.

Cost of Calcium Carbide. Engr., Dec. 22, 1899.

Cost of Producing Carbide by Different Agencies. F. Liebetanz. Jour. Gas Ltg., Dec. 26, 1890.

Crushing of Raw Materials for the Manufacture of Calcium Carbide. Elect'n, May 19, 1899.

Cubic Feet of Acetylene to a Pound of Carbide. H. F. Fuller. Dom. Eng., Jan., 1899.

Dangers of Acetylene. G. Jouane. Pro. Age, Feb. 15, 1898.
Dangers of Acetylene. Ad. Bouvier. Gas Wld., July 9, 1898.
Description of an Acetylene Gas Plant for the Great Southern and Western Railway. Gas Wld., Aug. 14, 1897.
Description of an Acetylene Light Plant. C. B. Orton. Gas Wld., Aug. 14, 1897.
Determination of the Heat Dissociation and of Combustion of Acetylene, Ethylene and Methane. W. G. Mixter. Am. Jour. Sci., Nov., 1901.
Determination of Phosphorus and Sulphur in Acetylene and Other Combustible Gases. P. Eitner and G. Keppler. Jour. Soc. Chem. Ind., page 938, 1901.
Development in Acetylene. J. H. Chewitt. Can. Eng., Aug., 1903.
Development of the Carbide Industry. Sci. Am. Sup., Aug. 13, 1898.
A Discussion of Certain Recent Data on the Cost of Acetylene and Calcium Carbide. W. H. Birchmore. Am. Gas Ltg. Jour., May 25, 1896.
Ueber diskontinuirliche und kontinuirliche Carbidöfen. O. Fröhlich. Zeit. Ei'ch., July 5, 1900.
Dissolved Acetylene. Engr., Oct. 20, 1905.
The Dissolved Acetylene Lighting System. Ry. Age, June 15, 1906.
Der Dritte Internationale Acetylen Congress in Paris, 1900. H. Koechlin, Oest. Woch., Dec. 21, 1901.
Dry Process of Generating Acetylene. Engr., Mar. 2, 1906.
l'Éclairage des Trains à Acétylène. Gén. Civ., July 3, 1897.
l'Éclairage des Voitures de Chemins de Fer. J. Carlier. Part I, Rev. Univ. d. Mines, May, 1902.
Economy of Compressed Acetylene Gas for Lighting Railway Cars, Stations and General Railway Light. E. G. Fisher. Pro. Iowa Ry. Club, Nov. 17, 1903.
De Erste Acetylene-Centrale in Nederland. J. A. De La Porte. Ing., June 15, 1901.
Eisenbahn-Waggon-Beleuchtung. H. Gerdes. Glas. Ann., Apr. 1, 1897.
Electro-Thermal Chemistry of Calcium Carbide. West. Elec., Nov. 16, 1895.
Ueber Elektrische Oefen, besonders mit Rücksicht auf die Darstellung von Calcium Carbid. B. Carlson. Serial. Zeit. El'ch., Feb. 8, 1900.
Elektrochemische Verhalten des Acetylens. H V. Steinwehr. Zeit. El'ch., June 6, 1901.

l'Étât Actuel de l'Industrie du Carbure de Calcium et de l'Acétylène. Oscar Münsterberg. Rev. Sci., June 15, 1902.

Experience with Acetylene Gas for Car Lighting. P. W. Resseman. Eng. News, Mar. 31, 1898.

Experiments with Acetylene Lamps for Pit-work. Abstract from Paper by G. Franke. Col. Guard., Mar. 21, 1902.

Experiments with the New Wolf Safety Lamp. G. Chesnau. Col. Guard., May 2, 1902.

Sur l'Explosibilité de l'Acétylène aux Basses Temperatures. G. Claude. Compt Rend., Jan. 30, 1899.

Explosibility of Liquid Acetylene. F. Clowes. Pro. Age, May 15, 1897.

Die Explosionsursachen von Acetylen. N. Caro. Verhandlungen des Vereins zur Beförderung des Gewerbfleisses, Berlin, 1906.

Explosion and Means of Preventing It. Sci. Am. Sup., Sept. 5, 1903.

Explosive Acts. Order in Council Relating to Acetylene Gas. London, 1897.

Explosive Power of Acetylene. Eng. News, Nov. 29, 1900.

Explosive Properties of Acetylene. F. C. Philips. Eng. Soc. W. Pa., May, 1897.

Explosive Side of Acetylene. F. H. McGahie. Sci. Am. Sup., Dec. 2, 1899, and Sci. Am., June 6, 1900.

Explosiveness of Acetylene Air Mixtures. H. Gerdes. Pro. Age, Dec. 1, 1898.

Fire Hazard of Acetylene. Am. Gas. Ltg. Jour., Apr. 27, 1896.

Fire Underwriters' Rules for Safe Introduction of Acetylene. Jour. Fr. Inst., Apr., 1899.

The First Calcic Carbide Plant. O. E. Dunlap. Min. & Sci. Press, Feb. 1, 1896.

First Production of Calcium Carbide in the United States. F. P. Venable. Dec. 16, 1898.

French Experiments on Acetylene and Alcohol as Motor Fuels. P. M. Heldt. H. Age, June 20, 1900.

Future of Acetylene. Holliday. Gas Wld., Apr. 17, 1897.

Future of Acetylene Lighting. G. L. Wilson. Met. Work, June 29, 1901.

Ueber den jetzigen Stand der Acetylen-Technik. Schw. Bauz., June H. Gerdes. Glas. Ann., Sept. 15, 1898.

Génerateurs Automatique d'Acétylène. Gén. Mod., Feb. 15, 1897.

Generation of Acetylen. C. A. S. Howlett. Cassier's, Mar., 1899.

Generation of Acetylene. J. M. Morehead. Electrochem. Ind., Oct., 1903.

Generation and Use of Acetylene. Bldr., Mar. 3, 1900.
Generation and Use of Acetylene. Sir Chas. Forbes. Jour. Gas Ltg., Apr. 17, 1906.
Generators for Acetylene. C. A. Hexamer. Jour. Fr. Inst., Apr., 1899.
Generators and Essentials of Safety. Stubbs. Engng., July 22, 1898.
German Chemists Discuss Acetylene. Gas Wld., Apr. 3, 1897.
A German Conference on Acetylene. Jour. Gas Ltg., Mar. 3, 1897.
Heating, Lighting and Ventilation of Cars. G. L. Fowler. R. R. Gaz., May 19, 1905.
Herstellungskosten von 1,000 kg. Calcium Carbid bei verschiedenen Betriebskräften. F. Liebetanz. Zeit. El'ch., Aug. 16, 1899.
Ueber den heutigen Stand der Acetylen und Carbid Industrie. V. Berdenich. Zeit. Oest. Ing., Apr. 13, 1900.
The History, Status and Possibilities of Acetylene. H. H. Suplee. Eng. Mag., Aug., 1897.
How Shall We Heat and Light Our Homes? M. & M. M. West. Lad. Home Jour., Feb., 1907.
How to Fit Up a Complete Acetylene Plant. F. Dye. Plumb. & Dec., Oct. 1, 1902.
Hygienic Aspect of Acetylene. Plumb. & Dec. Feb. 1, 1899.
The Illuminating Value of Acetylene. L. M. Bullier. Pro. Age.
Illumination of College Buildings by Acetylene. O. E. Dunlap. West. Elec., Aug. 21, 1897.
Illumination of Niagara Falls by Acetylene Gas. West. Elec., July 24, 1897.
Impurities and Purification of Acetylene. G. Keppeler. Sci. Am. Sup., Aug. 8, 1903.
Incandescent Acetylene Burners. F. H. Leeds. Jour. Gas. Ltg., June 19, 1906.
Incandescent and Carburetted Acetylene. N. Caro. Engr., Dec. 13, 1901.
Incandescent Lighting by Means of Acetylene and the Carburization of Acetylene. Sci. Am. Sup., Jan. 28, 1905.
l'Industrie de l'Acétylène en France de 1896 à 1905. Pierre Rosenberg. Paris, 1906.
Ueber den jetzigen Stand der Acetylen Technik. Schw. Bauz., June 10, 17 and 24, 1899.
Kosten der Calciumcarbid Darstellung. Elec. Tech., Apr. 30, 1898.
Kosten der Gebräulichsten Lichtquellen. Gesund. Ing., Mar. 15, 1898.

Lamp Black from Coal Gas and Acetylene. G. F. Jaubert. Pro. Age, Sept. 15, 1899.
Lecture upon Acetylene. J. M. Crafts. Sci., Mar. 13, 1896.
Letang and Serpollet's Acetylene Lamp. Sci. Am. Sup., May 1, 1897.
Light and Illumination. C. P. Steinmetz. Sci. Am. Sup., Mar. 30 and Apr. 6, 1907.
Lighting by Acetylene. T. Varney. Cassier's, Jan., 1901.
Lighting by Acetylene. F. Dye. Plum. & Dec., Aug. 1, 1901.
Lighting the Country House. E. D. Elliott. Count. Life, Oct., 1907.
Lighting with Dissolved Acetylene. Rev. Tech., Jan. 10, 1902.
Lighting of Railway Cars. G. D. Shepardson. West. Elec., May 18, 1901.
Lighting of Railway Cars. H. Gerdes. Pro. Age, Sept. 15, 1897.
London Acetylene Exhibition. Editorial. Pro. Age., Nov. 15, 1898.
Manufacture of Alcohol from Acetylene. Eng. News, May 31, 1900.
Manufacture of Calcium Carbide. J. T. Morehead and G. de Chalmont. Jour. Am. Chem. Soc., Apr., 1896.
Manufacture of Calcium Carbide. O. E. Dunlap. West. Elec., May 16, 1896.
Manufacture of Calcium Carbide. Abstract. M. Nicolai. Am. Mfr. Ir. Wld., Aug. 26, 1898, also Gas Wld., Sept. 10, 1898.
Manufacture of Calcium Carbide in Switzerland. E. A. Schneider. Ind. & Ir., Sept. 30, 1898.
Manufacture of Calcium Carbide. J. B. C. Kershaw. Elect'n, Nov. 23, 1900.
Manufacture of Calcium Carbide. H. Moissan. Eng. Mag., May, 1904.
Die Metall-Carbide. El'ch., Oct. 2, 1902.
The New Gaseous Illuminant, Acetylene. Ind. & East. Eng., Mar., 1900.
A New Plant for the Manufacturer of Calcium Carbide. J. C. Temple. Eng. News, Apr. 14, 1898.
The New Plant of the Union Carbide Co. at Sault Ste. Marie. Elec., Aug., 1898.
New Plant for the Union Carbide Company at Sault Ste. Marie, Mich. Elec. Wld., Aug., 1898.
Zur Nichtigkeitserklärung des D. R. P. No. 77168. Bullier Verfahren zur Darstellung von Erdalkalicarbiden. W. Borchers. Zeit. El'ch., Apr. 20 and May 18, 1899.
Le Noir d'Acetylene et des Derivés. E. Hubon, Mem. Soc. Ing. Civ., Nov. 12, 1900.

A Note on an Acetylene-in-Oxygen Flame. C. H. Sharp. Translation. Am. Inst. Elec. Eng., Dec., 1901.
Note sur l'Eclairage du Phare de Chassiron à l'Incandescence par l'Acétylène. de Joly. Ann. Ponts et Chaussés. 4th Trimestre. 1902.
Note on the Use of Acetylene Gas as an Illuminant for Polariscope Work. H. W. Wiley. Jour. Am. Chem. Soc., Feb., 1896.
Sur un Nouvel Etalon Lumineux. Ch. Féry. Compt. Rend., Apr. 25, 1898.
Nouvelle Méthode de Preparation des Carbures. Compt. Rend., Nov. 29, 1897.
Official Report of the Budapest International Acetylene Exhibition and Congress. Engr., Feb. 2, 1900.
L'Officina pel Carburo di Calcio di Milano. V. Lucchini. Elet. Dec. 1, 1900.
Oxy-Acetylene Lamp. Sci. Am. Sup., May 20, 1905.
Oxygen Flame. C. H. Sharp. Translation. Am. Inst. Elec. Eng., Dec., 1901.
Papers Read at the Acetylene Congress, as Taken from the Report of Herr Liebetanz. Serial. Jour. Gas Ltg., Oct. 17, 1899.
The Patent Side of Acetylene Gas. Pro. Age, July 15, 1897.
Petroleum Acts. Order in Council Relating to Calcium Carbide. London, 1897.
Physiological Action of Acetylene. W. H. Birchmore. Elec. Eng., Nov. 13, 1895.
Physiological Effects of Acetylene. T. Oliver. Pro. Age, June 1, 1898.
Poisonous Action of Acetylene. M. N. Gréhant. Elec. Eng., Dec. 11, 1895.
Poisonous Properties of Acetylene. R. Rosemann. Pro. Age, Mar. 1, 1896.
Portable Applications of Acetylene. Eng. Mag., Oct., 1903.
Position of Calcium Carbide and Acetylene. Engr., Sept. 13, 1901.
Possibilities of Acetylene. Am. Gas Ltg. Jour., Oct. 14, 1895.
Practical Use of Acetylene. Paper Read before Western Gas Association. Pro. Age, June 15, 1896.
Present Position of Acetylene Lighting. Paul Wolff. Jour. Gas Ltg., Nov. 8 and 15, 1898.
Present and Prospective Supply of Calcium Carbide. Cons. Rep., July 6, 1898.
Principle of Acetylene Lamps. S. W. Rushmore. Auto., Jan. 2, 1904.
The Principles of Illumination from the Standpoint of the Acetylene

Engineer. Van Rensselaer Lansingh. Read before the International Acetylene Association, Atlantic City, June 10, 1906.
Priority in Discovery of Commercial Method of Making Calcium Carbide. A. C. Fraser. Pro. Age, Feb. 1, 1898.
Priority in Electrical Process for Making Calcium Carbide. L. K. Boehm. Pro. Age, May 15, 1898.
Proceedings of the International Acetylene Association. Annual since 1898.
Production of Synthetic Alcohol. H. W. Wiley and H. Schreiber. Read before the Am. Phil. Soc., Apr. 20, 1907.
Products of the Explosion of Acetylene. W. G. Mixter. Am. Jour. Sci., Jan., 1900.
Progress in the Acetylene Gas Industry. J. B. C. Kershaw. Elec. Rev., Oct. 6, 1899.
Purification of Acetylene. Sci. Am., May 5, 1900.
The Purification of Acetylene. Jour. Gas Ltg., June 14, 1898.
Purification of Acetylene. Abstract. Jour. Gas Ltg., Dec. 30, 1902.
Purification of Acetylene by Means of Chloride of Lime. Engr., Nov. 3, 1899.
Purification and Combustion of Acetylene. Ullman. Sci. Am. Sup., June 18, 1904.
Quality of Calcium Carbide. Jour. Gas Ltg., May 7, 1898.
Railway Carriage Lighting. H. Gerdes. Gas Wld., May 1, 1897.
Railway Carriage Lighting by Compressed Oil Gas and Acetylene. Bork. Engr., Sept. 19, 1898.
Railway Carriage Lighting and the Latest Experiments with Acetylene. Ry. & Eng. Rev., Dec. 4, 1897.
Railway Lighting by Acetylene Gas. J. S. Seymour. Ry. & Eng. Rev., Aug. 24, 1901.
Rapport Général sur le Concurs d'Appareils à Acétylène, Paris, 1906.
Die Reaktiongleichung bei der Calcium-carbid-Reduktion. B. Neuman. Zeit. El'ch., Oct. 2, 1902.
Die Reaktionsgleichung bei der Calcium-carbid-Reduktion. Fr. von Kügelgen. Zeit. El'ch., Oct. 9, 1902.
Real Value of Acetylene as an Illuminant. Engr., Mar. 31, 1899.
Recent Facts about Acetylene. S. E. Filman. Cosmop., Nov., 1897.
Recent Research on Acetylene. G. E. Brown. Pro. Age, July 15, 1897.
Recherches sur les Dérivés Metalliques de l'Acétylène. Berthelot & Delépine. Compt. Rend., Aug. 14, 1899.
Regulations for Acetylene Lighting at Berlin. Translation. Pro. Age, Jan. 1, 1898.
Reply of Houston, Kennelly and Kinnicutt to Criticisms on the Pro.

Age. Report Concerning Cost of Carbide of Calcium. Pro. Age, July 15, 1896.

Reply of Pro. Age Commissioners to Criticisms on the Cost of the Carbide of Calcium. Pro. Age, July 15, 1896.

Report of the Acetylene Exhibition Committee. Engr., Mar. 3, 1899.

Report on Present Status of Manufacture of Carbide of Calcium and Acetylene Gas. Houston, Kennelly and Kinnicutt. Pro. Age, Apr. 15, 1896.

Report on the Use of Acetylene Gas by the Canadian Government as an Illuminant for Aids to Navigation. Albert Ross, Capt. U. S. N., Member of the Light House Board, Washington, D. C.

Restrictions placed on Carbide of Calcium in England. Pro. Age, Apr. 1, 1897.

Results of Practical Experiments with High Pressure Acetylene Lighting. H. T. Hauer. Pro. Age, July 15, 1901.

Risks in the Use of Calcium Carbide and Acetylene. E. V. French. Eng. News, Feb. 23, 1899.

Salvator Acetylene Searchlight. Sci. Am. Sup., Nov. 25, 1905.

Science of a Light. H. Maxim. Harp. Wkly., Jan. 23, 1904.

Serpollet Acetylene Generator. Sci. Am., Feb. 10, 1900.

Simple Generation of Acetylene and Its Purity. Jour. Gas Ltg., July 2, 1901.

A Sober Account of Acetylene. W. E. Gibbs. Jour. Gas Ltg., Aug. 23, 1898.

Solutions of Acetylene and Their Explosive Properties. Berthelot & Vieille. Pro. Age, Sept. 1, 1897.

Some Facts about Acetylene Gas. J. C. McMynn. Elec. Rev., Jan. 24, 1896.

Some Facts about the Purification of Acetylene Gas. Theo. Kautney. Dom. Eng., Sept. 15, 1900.

Some Enemies of Acetylene. Engr., Apr. 27, 1900.

Some More Enemies of Acetylene. Engr., June 1, 1900.

Some New Uses for Calcium Carbide. Engr., Jan. 18, 1901.

Some Notes on Calcium Carbide Manufacture. Elec. Rev., Lond., Sept. 20, 1901.

Some Properties of Acetylene Gas. Abstract. Ad. Bouvier. Sci. Am., Sept. 3, 1898.

Some Properties of Calcium Carbide. Venable and Clark. Am. Gas Ltg. Jour., page 528, 1895.

Studien über die Reduzierende Kraft des Calciumcarbids. Kügelgen. Zeit. El'ch., Apr. 11, 18 and 25, 1901.

Studies of the Explosive Properties of Acetylene. Berthelot & Vieille. Pro. Age, Dec. 1, 1896.
Symposium on Car Lighting. Ry. Age, Feb. 20, 1903.
Some Theoretical Aspects of Acetylene Manufacture. Engr., June 24, 1898.
La Soudure Autogéne des Metaux. André Binet. Gén. Civ., May 23, 1903.
La Soudure Autogéne des Metaux. P. Dumesnil, Mem. Soc. Ing. Civ., Nov., 1905.
Specifications for an Acetylene Gas Generating Apparatus for Fort McPherson, Ga. Eng. News, Feb. 4, 1904.
Storage of Calcium Carbide. Pro. Age, Mar. 1, 1897.
The Technical Analysis of Calcium Carbide and Acetylene and the Purifying of the Latter. Lunge and Cedercreutz. Pro. Age, Dec. 15, 1897, also Gas Wld., Jan. 1, 1898.
The Technology of Acetylene. Review of Lewes' Book. Jour. Gas Ltg., Aug. 21, 1900.
Three Phase Currents in the Manufacture of Calcium Carbide in Italy. Cesare Pio. Elec. Wld. & Elec. Eng., Aug. 5, 1899.
Town Lighting by Acetylene. Augustine Davis. Munic. Jour. & Eng., July, 1903; Oct., 1903; and Mar., 1904.
Town Lighting with Acetylene. Pro. Age, Dec. 1, 1898.
The Truth about Acetylene Gas. Albert Stetson. Elec., Feb. 12, 1896.
Two Systems of Using Acetylene Gas for Car Lighting. E. G. Fisher. Abstract. Eng. News, Oct. 1, 1903.
Usine à Gas Riche et à Acétylène. H. Guerin. Gén. Civ., Oct. 26, 1901.
The Utilization of Acetylene. Engr., Aug. 19, 1898.
Utilization of 110 Volt Electric Circuits for Small Furnace Work. Sci. Am. Sup., Aug. 27, 1898.
Ueber die Verbrennung des Kohlenstoffs bei der Calcium Carbid-Reduktion. Fr. v. Kügelgen and W. Halle. Zeit. El'ch., May 14, 1903.
Ueber Versuche mit Acetylen Beleuchtung in Bergwerken. G. Franke. Oest. Zeit. Berg. Hütt. Berlin, 1901.
Versuche mit der Wolf'schen Azetylen Sicherheitslampe. H. Beyling. Glück., July 8, 1905.
Velocity of Acetylene Detonations. Sci. Am. Sup., July 28, 1900.
Sur la Vitesse de Détonations de l'Acétylène. Berthelot and Le Chatelier. Compt. Rend., Aug. 28, 1899.
Welding by Acetylene. Ry. & Eng. Rev., Jan. 20, 1906.

Welding, a New System of. A. H. Sexton. Mech. Eng., June 17, 1905.
Zur Wertbestimmung elektrischer Kraft. O. Schmidt. Zeit. El'ch., Aug. 1, 1901.
Wet and Dry Generators. H. F. Fuller. Heat & Vent, Oct. 15, 1898.
Will there be a Carbide Industry? W. H. Birchmore. Eng. News, Jan. 30, 1896.
The Wilmington Explosion. W. A. Jones. Am. Gas Ltg. Jour., Jan. 24, 1898.
Die Wirkung der deutschen Arbeitversicherung. Gesund. Ing., Apr. 30 and May 15, 1897.
Die Zweite Internationale Acetylen-Ausstellung in Budapest, 1899. Karl Neudeck. Zeit. Oest. Ing., Jan. 12, 19 and 26, 1900.
Zweite Internationaler Congress for Carbid und Acetylen Industrie. H. Koechlin. Oest. Monat., Mar., 1900.

THE ACETYLENE PRESS.

The following are the principal journals especially devoted to the acetylene and calcium carbide industries:

United States.

The Acetylene Journal. Published monthly (48 pages) by the Acetylene Journal Publishing Co., 324 Dearborn St., Chicago, Ill. Edited by Elias A. Long; official organ of the International Acetylene Association; price fifty cents per year; now in its tenth year.

England.

Acetylene. Published monthly (24 pages) at 46 Southampton Buildings, Chancery Lane, London, W. C. Edited by F. H. Leeds; official organ of the Acetylene Association; price three shillings per year; now in its sixth year. This journal replaces an earlier periodical known as the Journal of Acetylene Lighting, which was issued monthly from 1898 to 1902.

Germany.

Zeitschrift für Calciumcarbid-Fabrikation, Acetylen-und-Klein-Beleuchtung. Published weekly (8 pages) at Kommandantenstr. 14, Berlin S. W. Edited by J. H. Vogel; official organ of the German Acetylene Association; price sixteen marks per year; now in its twelfth year.

Acetylen in Wissenschaft und Industrie. Published semi-monthly (12 pages) by Carl Marhold, Halle. Edited by Theodor Kautny; price sixteen marks per year; now in its eleventh year.

France.

Revue des Éclairages. Published semi-monthly (12 to 16 pages) at 104 Boulevard de Clichy, Paris. Edited by R. Granjon; price eight francs per year; now in its thirteenth year.

Bulletin Officiel de l'Union des Propriétaires d'Appareils à Acétylène. Published monthly at 104 Boulevard de Clichy, Paris. Price three and a half francs per year.

Spotted something that isn't quite right?

Let us know and we'll fix it for future reprints. And as a thank you for your diligence, we'll send you a free Sothis book of your choice.

publisher@sothis.press

This page intentionally left blank

Made in United States
Orlando, FL
06 November 2026

90301612R00074